About Island Press

Since 1984, the nonprofit organization Island Press has been stimulating, shaping, and communicating ideas that are essential for solving environmental problems worldwide. With more than 1,000 titles in print and some 30 new releases each year, we are the nation's leading publisher on environmental issues. We identify innovative thinkers and emerging trends in the environmental field. We work with world-renowned experts and authors to develop cross-disciplinary solutions to environmental challenges.

Island Press designs and executes educational campaigns, in conjunction with our authors, to communicate their critical messages in print, in person, and online using the latest technologies, innovative programs, and the media. Our goal is to reach targeted audiences—scientists, policy makers, environmental advocates, urban planners, the media, and concerned citizens—with information that can be used to create the framework for long-term ecological health and human well-being.

Island Press gratefully acknowledges major support from The Bobolink Foundation, Caldera Foundation, The Curtis and Edith Munson Foundation, The Forrest C. and Frances H. Lattner Foundation, The JPB Foundation, The Kresge Foundation, The Summit Charitable Foundation, Inc., and many other generous organizations and individuals.

The opinions expressed in this book are those of the author(s) and do not necessarily reflect the views of our supporters.

Arctic Passages

Arctic Passages

ICE, EXPLORATION, AND THE BATTLE
FOR POWER AT THE TOP OF THE WORLD

Kieran Mulvaney

⬤ **ISLAND**PRESS | Washington | Covelo

Design and Typesetting by 2K/DENMARK.

Sustainable Typesetting® ID: ST-49535052-24-A01
The ID number certifies that this book fulfills the standards and recommendations of Sustainable Typesetting®, providing reduced CO_2 emissions.

Library of Congress Control number: 2024945776

All Island Press books are printed on environmentally responsible materials.

Manufactured in the United States of America
10 9 8 7 6 5 4 3 2 1

Generous support for this publication was provided by the Curtis & Edith Munson Foundation.

Keywords: Arctic, Arctic Ocean, sea ice, climate change, Russia, China, Canada, geopolitics, exploration, Northwest Passage, Northeast Passage, Northern Sea Route, North Pole, navigation, shipping

To my brothers

Contents

Author's Note
and Acknowledgments

In January 1993, I stood on the deck of a ship in Antarctica and gazed in wonder at the frozen continent's largest active volcano, shrouded with snow and adorned by glaciers and defiantly venting steam into the freezing air. I was on the second of what would be four expeditions to Antarctica over the course of a decade, but the sight of that volcano set in motion a love affair with the other polar region.

The mountain was first seen in January 1841 by the members of a voyage of discovery led by British explorer Sir James Clark Ross, their sense of wonder heightened by the fact that their arrival coincided with its being in the middle of an eruption. Marveled Joseph Hooker, the ship's surgeon: "This was a sight so surpassing everything that can be imagined ... that it really caused a feeling of awe to steal over us."

Ross dubbed the volcano Mount Erebus and its extinct companion Mount Terror, after the expedition's two ships. Following the success of Ross's endeavor, HMS *Erebus* and HMS *Terror* combined for another voyage of discovery, this time in the Arctic as they sailed under the command of Sir John Franklin in search of the fabled Northwest Passage.

They would never return.

The disappearance of *Erebus* and *Terror* set in motion a series of search attempts, several of which resulted in the ships of the would-be rescuers also becoming trapped in the ice of the Canadian Arctic. It birthed a cottage industry of speculation as to their fate and romanticism over their heroic effort and glorious failure. The Franklin expedition became a byword for Victorian Arctic exploration and an integral part of the Arctic mythos, and thus national identity, of Canada. My first sight of Mount Erebus prompted an ongoing interest in the ships that gave it and its twin their names, the voyages they undertook, and in particular the part of the world in which they met their mysterious demise. That process ultimately manifested in the book you are holding in your hands right now.

My initial intent was a largely historical tome; I then contemplated a more straightforward analysis of the present geopolitical situation surrounding the melting Arctic and in particular the opening of the Northwest, Northeast, and Transpolar Passages. The ultimate result is something of a hybrid: an examination of the contemporary situation placed within a historical context, the present interest in taking advantage of an ice-free Arctic contrasted with the fates of expeditions that were destroyed, one after another, by sea ice that now is in retreat. I have also striven to provide a sense of place, an understanding of what the Arctic is and of the importance of sea ice for the people and wildlife that inhabit the region, as well as for the planet as a whole.

Circumstances conspired to make writing this book a far more challenging experience than anticipated. The narrative is built around a 2017 voyage to the North Pole onboard the Russian icebreaker *50 Let Pobedy*; a 2019 journey through the bulk of the Northwest Passage on the *Ocean Endeavour*, operated by Adventure Canada; and a 1998 trip to the Alaskan and Russian Arctic that I made as a journalist onboard the Greenpeace ship MV *Arctic Sunrise*. I had planned a further voyage, through the Northeast Passage, but as plans began to take fruition, they were upended by the COVID-19 pandemic and then Russia's invasion of Ukraine.

That forced something of a scramble as I reoriented parts of the book, but the lingering effects of the pandemic, and in particular a severe mental health crisis that left me feeling adrift on an ocean of depression, slowed writing down to a crawl. Fortunately, that crisis has passed, even as the one facing the Arctic persists.

I would like to express my deepest, undying thanks to my editor at Island Press, Erin Johnson, who fought for the book and alternately encouraged and cajoled me, steering me through my downtimes and into a period of productivity that brought the book to its conclusion. I am eternally indebted also to my agent, John Thornton, for his tireless, gentle, and unflappable support from first pitch to final publication.

Many other thanks are due:

To Arne Sorensen and Bob Graham, my first polar captains, who taught me so much about the Arctic and Antarctic;

To my friends and colleagues at Polar Bears International, notably Barbara Nielsen, Krista Wright, and Geoff York, for their continued support, friendship, and opportunities;

And to those who helped me so much with opinions, information, and interviews during the writing of this book, notably Dmitry Lobusov and Vladimir Yudin, respectively the captain and chief engineer onboard *50 Let Pobedy*; Solan Jensen and Colin Souness of Quark Expeditions; Jason

Edmunds, expedition leader onboard *Ocean Endeavour*; Victoria Polsoni and Latonia Hartery of Adventure Canada, who were such helpful companions during the Northwest Passage voyage; Erin Morawetz, who made my presence on that voyage possible; Brandy Lockhart and Tamara Tarasoff of Parks Canada; Mia Bennett of the University of Washington; Marisol Maddox of the Polar Institute at the Woodrow Wilson International Center for Scholars; Melody Schreiber of *Arctic Today*; Mark Serreze of the National Snow and Ice Data Center; David "Duke" Snider of Martech Polar Consulting; Mike Spence; Ian Van Nest; Paul Ruzycki; Larry Audlaluk; Anthony Soolook Jr.; Jesse Ningiuk; Matthildur María Rafnsdóttir; and Ólafur Ragnar Grímsson, former president of Iceland and now chair of Arctic Circle.

(It goes without saying, but should be articulated anyway, that the help of the foregoing does not necessarily translate to an endorsement on their part of what I have written; any emphases, implications, and errors are mine and mine alone.)

Portions of this book were first published, in different forms, by Discovery Channel News, *National Geographic*, the *Guardian*, and the *Washington Post Magazine*; many thanks to my editors Lori Cuthbert, Sarah Gibbens, Jessica Reed, and David Rowell for their support, encouragement, and commissions. (It is a sad testament to the state of the modern media industry that only two of those titles remain extant.)

My deepest gratitude to the Curtis & Edith Munson Foundation for providing additional funding that made this book possible; many thanks to Angel Braestrup for steering that funding through. I hope the final result is to your satisfaction.

Book writing can be a lonely and isolating process, and I am more grateful than I can ever say for those who make my life so much better, particularly Rachel Charles, Loni Laurent, Sam Walton, Melanie Duchin, Sallie Schullinger, Deb and Jim Cossaart, Chris Kolliander, the late Earle Ray, Piper Wallace Westbrook, Sara Giard, and, above all, Sarah Jean Luke, who has lived and suffered through every moment of this book alongside me.

A special meow to Alfie, Zoe, Natalia, and Lucy.

Finally, my ongoing love to my family, especially my brothers, Michael and Stephen, to whom this book is dedicated.

Kieran Mulvaney
Bristol, Vermont
July 18, 2024

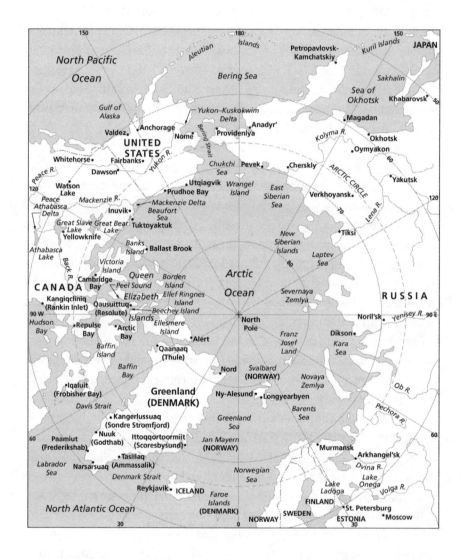

Prologue: Beginnings

The *Ever Given* is among the largest oceangoing vessels ever built, a giant floating slab of steel that appears, to use one memorable description, "more sideways sidescraper than boat." More than 1,300 feet long and almost 200 feet wide, it was en route from Tanjung Pelepas in Malaysia to Rotterdam, stacked high with containers full of furniture and pillowcases, bicycles and barbecues, swimwear, microchips, sex toys, and myriad other vital components of twenty-first-century Western living, when, in the early hours of March 23, 2021, it steamed north into the Suez Canal.

That stage of its voyage had barely begun when, at approximately 7:40 a.m. local time, a fierce windstorm swept out of the desert. Gusts of close to fifty miles per hour buffeted the *Ever Given*'s fourteen stories of containers, which effectively acted like a giant sail, pushing the ship toward the canal's edge. In the wheelhouse, the pilots argued with each other and with the ship's captain over the appropriate corrective actions, countermanding each other's orders as the vessel whipsawed in the wind until the bow ground into the canal's eastern bank and its stern swung outward. Such was the vessel's size that by the time the storm abated, the *Ever Given* had completely blocked the canal.

At that moment, 15 vessels were stuck behind it in the northward queue. Five days later, with the *Ever Given* still a day away from being freed, at least 369 ships were waiting to pass through the canal from either end, including 41 bulk carriers, 24 crude oil tankers, 2 Russian navy vessels, and the *Ever Given*'s sister ship, the *Ever Greet*. The blockage added further delays and price increases to a global supply chain that was already bursting at the seams as a result of the unprecedented stresses caused by a global pandemic; *Lloyd's List* estimated that the shipping industry incurred losses in the region of $400 million every day that the *Ever Given* remained in situ. Egypt announced afterward that it would widen the part of the canal in which the *Ever Given* had become a temporary resident, much as it had previously widened a more northerly segment of the canal to encourage more ship traffic. That previous widening, however, had not had the desired effect: instead of more vessels, it encouraged larger ones, of which the *Ever Given* was but one example. In the accident's aftermath, industry postmortems focused on the growing reliance on a smaller number of larger ships and the concomitant risks that entailed; but from points farther north came mischievous whispers that a greater problem was the fact that so much of the world's transoceanic shipping trade is funneled through two narrow bottlenecks—the Suez and Panama Canals—and that what was really needed was the development of alternative routes. Specifically, those voices reasoned, the future of global shipping lay in the Arctic.

On its face, the notion of a major commercial shipping lane in the Arctic seems absurd. This is, after all, a part of the world that was described by early twentieth-century Arctic missionary S. K. Hutton as a "bare and desolate waste, silent but . . . for the dismal howling of the hungry wolf, or the even more dismal howling of the wind," its constituent parts dismissed by Charlotte Brontë in *Jane Eyre* as "those forlorn regions of dreary space," a region whose very name is used as a descriptor of bitter cold. And indeed, the Arctic can be staggeringly frigid: the lowest temperature ever recorded in the Northern Hemisphere was a scarcely comprehensible –93.3°F, recorded atop the Greenland ice sheet in December 1991; even in the central Siberian Arctic, the average winter temperature is between –30°F and –35°F. In Utqiagvik, Alaska, the northernmost town in the United States, the Sun disappears below the horizon in November, not to appear again until January—and even then, initially for only fifty minutes.

There are a number of ways to define the Arctic, all of which serve to emphasize the region's harsh environment. They include the tree line— north of which forest yields to tundra—and its marine equivalent, the southern limit of winter pack ice. Another is the 10°C (50°F) isotherm, a line that delineates the average temperature during the warmest month of the year—north of that line is the Arctic; immediately south of it is the subarctic. The Arctic Circle, which encircles the globe at latitude 66°30'N, marks the point above which the Sun does not set on the summer solstice or rise on its winter equivalent, but it is not an ecologically useful delineation.

The entire Arctic, depending on how it is defined, covers an area of approximately 11–14.5 million square miles of land and sea, and it is permanently inhabited by no more than four million people, less than half the population of New York City in an area forty-five thousand times as large. Nunavut, the largest territory in Canada—which is itself the second-largest country in the world—is home to a little fewer than 40,000, and its most populous municipality, Iqaluit, boasts roughly 7,500 residents.

But the Arctic is more than cold and ice and empty spaces. As early twentieth-century explorer Vilhjalmur Stefansson, perhaps the region's most evangelical nonresident advocate, sarcastically observed, the region "is lifeless, except for millions of caribou and foxes, tens of thousands of wolves and muskoxen, thousands of polar bears, millions of birds, and billions of insects." And few as the region's inhabitants may be, some of them live in towns and cities that compare favorably in size with those in any number of more salubrious climes: 75,000 or so in Tromsø in Norway; 178,000 in Norilsk in Siberia; and almost 300,000 in the region's biggest city, Murmansk, on Russia's Kola Peninsula—roughly as many people as live in Saint Louis, Missouri, or Venice, Italy. And yes, there are commercial sea routes, including one that lies entirely above the Arctic Circle that sweeps farther north than any point in Alaska or Scandinavia and carves a path north of Russia. In its entirety, from the Bering Strait in the east to the Barents Sea in the west, it is known as the Northeast Passage; to Russia, the part of the journey that encompasses its national waters—and which comprises 90 percent of the passage's full extent—it is the Northern Sea Route (NSR). And it is that route that Russian authorities advanced as an alternative to the Suez Canal as crews struggled to free the *Ever Given*.

"Obviously, it's necessary to think about how to efficiently manage transportation risks and develop alternative routes to the Suez Canal, first and foremost the Northern Sea Route," Nikolai Korchunov, Moscow's ambassador-at-large, told the Interfax news agency. "The appeal of the Northern Sea Route will grow both in the short- and long-term."

The development of the NSR has been decades in the making. After centuries of exploration and tentative commercial activity along parts of the northern Russian coast, the Soviet Union in 1932 formed the Directorate of the Northern Sea Route to develop the passage as a viable transport route. Two years later, the icebreaker *Fyodor Litke* became the first vessel to transit the entire route, east to west, in one season, and two years after that it returned to escort the first two warships to transit the passage, the *Voykov* and the *Stalin*.

Over the subsequent ninety years, the NSR has come to assume an oversize importance in the collective Russian psyche, an integral symbol of a prosperous Arctic future. And that importance has been reflected in growing governmental investment, particularly under the reign of Vladimir Putin. In 2021, almost 35 million tons of cargo traveled along the route, a fifteenfold increase over the preceding decade; a 2018 decree from Putin set targets of 80 million tons by 2024 and 110 million by 2030. The number of voyages involved in transporting that much cargo also has been increasing: from 18 in 2015 to 85 in 2021. But while such figures may seem impressive out of context, and are certainly notable, they also show just how far away the NSR is from challenging the Panama and Suez Canals, which average approximately twelve thousand and eighteen thousand transits each year, respectively.

That said, the NSR is orders of magnitude more developed as a transit route than another putative passageway through the Arctic that has, if anything, been even more celebrated for its potential, even as it has proven more resistant to human desires. Starting, like its Northeast cousin, in the Bering Strait, and threading its way through Canada's Arctic islands to the Atlantic Ocean, the Northwest Passage occupies a singular place in the mythologies of Canada and its former colonizer, Britain, the result of centuries of frequently desperate attempts to probe its secrets, attempts that time and again foundered in the face of opposition from unyielding Arctic ice.

Historical interest in finding such a pathway was understandable. Prior to the opening of Suez in 1869, the only available options for transoceanic travel involved sailing south of Africa or, yet more treacherously, South America—options that, at various points in history, presented as much of a geopolitical challenge as a navigational one.

That such interest endures owes much to the lingering romantic attachment to the *idea* of a Northwest or Northeast Passage, the notion that anything that has involved so much sacrifice and effort over so many generations should not be too easily cast aside. It is testament, too, to the tightness with which Canada and Russia cling to their image as Arctic superpowers and the role that those prospective passageways may yet play in manifesting that image. It is a result also of cold calculation, of the recognition that an unimpeded journey through the Northwest Passage from, say, Rotterdam to Yokohama would be approximately 4,500 miles shorter than one via the Panama Canal, or that a similar voyage through the Northeast Passage would be roughly 40 percent shorter than traveling through the Suez Canal.

But the reason Putin is able to proclaim that "the Arctic is the shortcut between the largest markets of Europe and the Asia-Pacific region" with any kind of seriousness, or that an editorial on Al Arabiya can lament, in the shadow of the *Ever Given* incident, that navigating the Northwest Passage is "getting closer" and that it "will eventually replace the Suez Canal as the world's favored shipping route to link the Atlantic and Pacific Oceans," is that circumstances are not what they were when Victorian explorers were dying on the ice that crushed their ships.

The climate is changing, the world is warming, and the Arctic is melting. As a result, while the Northwest and Northeast Passages are not yet reliably navigable, they are at least somewhat and seasonally so. The Northwest Passage, which had not been transited at all prior to 1906, and had been so only four times by 1954, saw twenty-five transits in 2021 alone—a consequence of sturdier ships and a stronger commercial imperative, certainly, but also of the fact that, at least for a couple of months per year, navigating Canada's Arctic waters is significantly less hazardous than had long been the case. That is why former US secretary of state Mike Pompeo, chief diplomat of an administration that did not even recognize the scientific realities of global warming, nonetheless argued in 2019 that "steady reductions in sea ice are opening new passageways and new opportunities for trade. This could potentially slash the time it takes to travel between Asia and the West by as much as 20 days. Arctic sea lanes could [become] the 21st-century Suez and Panama Canals." It is a prospect that has countries other than Canada and Russia, friends and foes alike, watching with anticipation. It has helped ignite a simmering and thus far low-grade dispute between Canada and the United States. And as China has grown in economic might and geopolitical heft, it too has begun exploring the expansion of its naval and commercial reach by investing in its own fleet of icebreakers, seeking a seat at the table of Arctic nations, and exploring the prospect of itself taking advantage of new pathways through the ice—through North America, across the top of Russia, or even via the North Pole.

There is a certain circularity about China's involvement in the search for a passage through Arctic ice, since it was to reach China that European nations began that search more than half a millennium ago.

—————∞—————

The two small ships that sailed down the River Thames early in the summer of 1576 were on a special mission. The *Michael* and the *Gabriel* were setting out on an expedition into the unknown, the import of which was underlined by the fact that as they passed the royal palace in Greenwich, Queen Elizabeth I herself waved from her window. The two ships, under the command of Martin Frobisher, were on a quest to find a seaway from the Atlantic to the Pacific across the top of North America. Frobisher was convinced that, just as Ferdinand Magellan had found a route between the two oceans to the south of the Americas, he would uncover a more northerly path, "that such a passage was as plausible as the English Channel."

Such optimism proved misplaced. Frobisher reached the west coast of Greenland and the easternmost extent of the Canadian Arctic, lost five men who went ashore but did not return, and kidnapped an Inuk, who caught a cold on the voyage home and died shortly after the ships arrived back in England. Frobisher did not, however, find the Northwest Passage.

His desire to do so was anchored not in a desire to explore for exploration's sake but in a series of developments that had unfolded in short order at the tail end of the previous century, a decade described by historian Felipe Fernández-Armesto as the "great leap forward" in western European exploration. To understand those developments, we begin with Christopher Columbus.

When Columbus set sail from Palos de la Frontera in August 1492, it was with the intention of finding a route across the Atlantic to Asia, specifically China. The area enticed with promises of spices and gold, but land routes from Europe to the region were long and challenging and frequently dominated by Muslim traders, while the Portuguese had developed and monopolized a sea route to the east around Africa. Columbus, born in Genoa but soliciting support from Portugal's Spanish rivals, proposed sailing west, in the process transforming the Atlantic from a frontier precluding Europe's westward expansion into a corridor facilitating it. He was not the first to give voice to such an idea; but the notion had previously failed to gain traction because of the consensus that Asia was simply too distant for such a voyage to be plausible.

Contrary to lore, neither Columbus nor his contemporaries believed the world to be flat. He did, however—whether out of conviction or convenience—argue that it was smaller than was generally agreed. Around 500 BCE, Pythagoras had proposed that Earth was essentially spherical, and approximately 180 years later, Aristotle proffered evidentiary documentation: constellations change their apparent position in the sky as a traveler moves north or south; Earth casts a circular shadow on the Moon during a lunar eclipse; ships disappear hull first when they sail over the horizon. In approximately 220 BCE, Eratosthenes went one further, using shadows cast by sticks at noon to produce an estimate of the circumference of the spherical Earth that was within spitting distance of what we now know to be the correct value of 24,900 miles at the equator. Columbus, however, promoted a fringe theory that suggested the planet was as much as 20 percent smaller than the figures of Eratosthenes and reality. On top of that, he argued, China extended farther eastward than had been previously recognized; accordingly, he insisted, it would be possible to reach Asia "in a few days."

It in fact took him a couple of months to make landfall, on October 12, 1492, probably on what is now known as San Salvador Island in the Bahamas. Although he spent a great deal of effort trying to establish otherwise, it is not a matter of dispute that neither the Bahamas nor Cuba, where he landed twelve days later, nor Hispaniola, which he reached on December 4, is in Asia; nor does it appear that, upon his return, many bought his sometimes convoluted assertions to the contrary. Indeed, it was the recognition that Columbus had in fact stumbled across lands hitherto entirely unknown to Europeans that prompted a new wave of exploration, motivated by trade, colonization, religious conversion, and combinations of these—as well as, ironically, a recognition that Columbus's voyages revealed there were staging posts in the ocean that could be used for rest and resupply en route to Asia.

Two years after Columbus's voyage, and in response to a growing diplomatic contretemps over his pronounced discoveries, Spain and Portugal signed the Treaty of Tordesillas, which effectively divided up new lands between them via a line of demarcation 370 leagues (approximately 345 miles) west of the Cape Verde Islands—roughly longitude 46 degrees 30 minutes west. Everything west of that line would be Spain's; anything to the east, Portugal's. That effectively granted Spain the right to the Americas (except for the as-yet-unknown-to-the-signatories eastern part of Brazil, which fell on the Portuguese side of the line) and ceded Africa to Portugal. The feelings and opinions of the people who already resided on those continents were neither considered nor acknowledged; the only concession of sorts to preexisting human presence was the caveat that any discovered nations with a "Christian King" would not be colonized.

No other country recognized the treaty, but such was the dominance of the Iberian nations' fleets that it soon assumed a force of de facto recognition, if not one of de jure. An initial wave of post-Columbian expeditions had not yet established the continental nature of the Americas, encouraging continued searching for routes to the east via the west, through and around these newly discovered lands; denied footholds to the south as a result of a rapidly growing Spanish and Portuguese presence, other nations looked to find routes to the north.

In 1497, John Cabot, a Venetian citizen, left the English port city of Bristol with "one ship of fifty toneless [tons] and twenty men and food for seven or eight months," carrying a commission from England's Henry VII and a remit to pursue a more northerly course than had Columbus. On June 24, he and a small party set foot on what was probably Newfoundland but may have been Cape Breton Island or possibly Nova Scotia. Unlike Columbus, he had reached the North American continent, but like him, he seems to have believed—or at least proclaimed—that he had made it to Asia, declaring his landing spot to be "the country of the Great Khan." King Henry awarded him the sum of £10 for his labors, and the following year, Cabot set out again, this time at the head of an expeditionary force of five ships. Its fate is uncertain; one of the five ships apparently became separated from the rest and limped, storm damaged, into port in Ireland. Recent research indicates that Cabot may have been alive and well in England in 1500 and suggests not only that the rest of his fleet did reach Newfoundland, but also that some members of his expedition may have founded a community, including North America's first Christian church, there. But details of the voyage and its aftermath are sketchy at best, and for several centuries it was assumed that all had perished at sea. Cabot was widely believed, in the words of one contemporary author, to have found the new lands "nowhere but on the very bottom of the ocean, to which he is thought to have descended together with his boat, the victim himself of that self-same ocean; since after the voyage he was never seen again anywhere."

Even if Cabot did perish, his influence and involvement did not. In 1499, William Weston, a Bristol merchant who appears to have been an associate of Cabot, set sail, according to correspondence authored by the king, "for to serche and fynde if he can the new founde land." The late historical researcher Alwyn Ruddock, who was the first to suggest Cabot had not perished on his 1498 voyage, believed that Weston returned to Newfoundland and continued north as far as the entrance to what we now know as Hudson Bay, a journey that would have taken him into ice-covered waters. "The only plausible explanation" for such a route, opined Evan Jones and Margaret Condon of the University of Bristol's Cabot Project, "was that the explorers were looking for a northern route around the landmass." If so, they noted, it was not only the first English-led expedition to America; it was also the first journey undertaken specifically to find not new land but a passage through and around that land to Asia. It would not be the last.

Cabot's son Sebastian may have participated in or led as many as three expeditions in the first decade of the sixteenth century, the last of which, conducted from 1508 to 1509, reportedly consisted of two ships and three hundred men and reached as far north as Hudson Bay and then as far south as Chesapeake Bay, close to what is now Washington, DC. By the middle of that century, he was Governor for Life of the Muscovy Company, granted a monopoly by Elizabeth I on trade with Russia and areas to the north. The Muscovy Company evolved from the Mysterie and Company of Merchant Adventurers for the Discoverie of Regions, Dominions, Islands and Places Unknown, whose first venture was in search of an alternative route to Cathay (China and environs): a North*east* Passage, across the top of Eurasia. That venture saw three ships—the *Edward Bonaventure*, the *Bona Confidentia*, and the *Bona Esperenza*—depart England in May 1553; only the *Edward* returned. The other two were swept east into what is now the Barents Sea during a storm and were forced to spend a winter on the Murman Coast of the Kola Peninsula, where their combined crews of approximately seventy officers and men "perished miserably from the effects of cold, or hunger, or both." The surviving vessel made it as far as the site of present-day Arkhangel'sk, on the coast of the White Sea, where its appearance initially terrified local fishermen, who threw themselves to the ground and tried to kiss the feet of the captain, Richard Chancellor. Chancellor learned from them that the country was called "Muscovy, which hath the name also of Russia the White," and that it was ruled by Ivan Vasilyevich—who would later become known as Ivan the Terrible and who, upon being advised of "the arrival of a strange nation," summoned Chancellor to the country's capital, Moscow, where he presented him with a letter to the king that granted freedom of trade to English ships.

The following spring, Chancellor returned to his ship and sailed home, where his news of a new market to the east was received with rapturous enthusiasm. Tragically, he perished on his return from a second voyage in a fierce storm that also finally claimed the *Bona Esperenza* and the *Bona Confidentia*, which were being sailed back to England by new crews from their enforced sojourns on the Kola Peninsula.

Where the English went, the Dutch—newly expansive after proclaiming independence from Spain in 1581—followed. Between 1594 and 1597, three successive voyages, led by Willem Barents and Jacob van Heemskerck, sought to find a route to Cathay by rounding the Russian Arctic archipelago of Novaya Zemlya. The voyages were not without success: they charted the coast of Novaya Zemlya and made the first confirmed sighting of Spitsbergen (although Barents and van Heemskerck thought it to be part of Greenland). But the first two voyages were halted by ice in the Kara Sea, to the east of

Novaya Zemlya, and the third saw Barents's ship entombed in ice and the crew forced to spend the winter of 1596–97 marooned ashore, barricading themselves against predatory polar bears.

While the primary focus was on putative passages to the northwest and northeast, there were some efforts to explore the potential of a third pathway: a route through the Arctic Ocean via the North Pole. The suggestion had been made by, among others, Bristol merchant Robert Thorne, who in 1527 declared in response to skeptics that "there is no land uninhabitable, nor sea unnavigable." The failure of expedition after expedition challenged the credibility of his assertion.

In 1607, Henry Hudson—about whom little is known prior to this commission—was dispatched by the Muscovy Company in search of Thorne's putative polar passage. After passing close to the east coast of Greenland, he turned northeast, spied Spitsbergen, and reached about 80 degrees North (80°N) before, being faced by an impenetrable sea of ice, turning south. The following year, the Muscovy Company sent him in search of the Northeast Passage; he explored and landed on the coast of Novaya Zemlya, attempted and failed to pass south of the archipelago, finding his way again blocked by ice, and then turned west in search of a nonexistent "Willoughby's Land." His achievements had fallen far short of what his benefactors hoped for, and with his return the Muscovy Company opted out of funding any more efforts to find a way through the ice of the eastern Arctic.

Hudson turned to the Dutch—specifically, the Amsterdam chapter of the Dutch East India Company, which sponsored his return to the Russian Arctic in 1609. Once again, he searched for a route south of Novaya Zemlya; once again, he found his way blocked by ice. This time—and, alas for Hudson, not for the final time in his career—his crew threatened to mutiny if he persisted in his quest to push through the ice, and so he turned west and sailed across the Atlantic in pursuit of the Northwest Passage. In early September, he reached the mouth of a mighty river that now bears his name; the following year, he returned, on what would be his final journey. Passing through what is now Hudson Strait, he found himself in what at first appeared to be open ocean, which he assumed to be the Pacific. But he had not completed the Northwest Passage; what he thought was an ocean was in fact a largely enclosed body of water now known as Hudson Bay, in which his ship became entrapped by ice over winter. That winter was not a pleasant one: as food supplies diminished, some of the crew became fractious and even deeply hostile, believing that Hudson was hoarding rations for himself and his favorites.

By the time the ice had released its grip and the ship prepared to leave for England, the situation onboard had devolved into open rebellion: on June 23, 1611, mutineers placed Hudson, his teenage son, and seven crew members who had scurvy or who were considered loyal to Hudson into a small boat and lowered them over the side to their fate, never to be seen again.

Upon their return, the surviving eight crew members faced remarkably little sanction for their actions: partly, probably, because they made a point of affixing the blame for organizing the revolt on crewmates who had perished on the journey home and were thus unable to offer contrary narratives; and partly, possibly, because the knowledge of the area they brought back with them rendered them too valuable to rot in incarceration. Of foremost interest among the information that they reported was Hudson's belief that they had found the Northwest Passage, which sparked a final, dying ember of voyages seeking either affirmation or refutation.

None of these voyages found a Northwest Passage, any more than Chancellor, Barents, or the Muscovy Company had succeeded in navigating a path to the Northeast. And yet the expeditioners of the seventeenth century had come closer to uncovering the secret entryway to the Canadian Arctic than anyone realized. On March 26, 1616, the *Discovery*—the same ship from which Henry Hudson had been expelled four years previously—set sail from Gravesend on the River Thames under the command of Robert Bylot (one of the crew who had set Hudson adrift) and with William Baffin as navigator. Their orders were to retrace and expand upon a trio of voyages led by John Davis between 1585 and 1587, who at his northernmost point had sailed north of 72°N; had encountered whales, Inuit, and "white beares of monstrous bignesse"; had painstakingly mapped the coastal areas he encountered; and had declared confidently that "it appeares that from England there is a short and speedie passage into the South Seas, to China, Molucca, Philippina, and India, by northerly navigation." To this end, they entered the strait between southwestern Greenland and eastern Canada that now bears their predecessor's name, sighting Greenland in mid-May and continuing northward up its western coast. On May 30, they passed the most northerly point reached by Davis and entered waters subsequently known as Baffin Bay, which had not been seen by Europeans since Viking times. On July 5, they reached the northernmost point of their expedition before swinging about and sailing down the west side of Baffin Bay, along the eastern coast of northernmost Canada. Here, they identified and named Lancaster Sound, which we now know is the primary entryway to the Northwest Passage; but the ship's crew was by now beginning to succumb to the crippling embrace of scurvy, and the priority of Baffin and Bylot was to set ashore at a place where they could procure curative herbs. Of Lancaster Sound, Baffin noted little other than that its entrance was full of ice. The prospects of a Northwest Passage, he concluded, appeared grim.

Even so, his observations might have encouraged further examination had they been made widely available; in the event, however, they fell victim to an ill-timed misfortune. Previous English Arctic voyages had been compiled and chronicled by Richard Hakluyt, writer and evangelical proponent of his country's colonization of North America, as part of his three-volume work *The Principal Navigations, Voiages, Traffiques and Discoueries of the English Nation*, published in 1589 and greatly expanded between 1598 and 1600. He continued his work after the publication of his magnum opus, but he died just three months after the *Discovery* returned to England; the man who assumed his responsibilities, the Reverend Samuel Purchas, chose to include only a truncated narrative in his successor work, *Hakluytus Posthumus, or Purchas his Pilgrimes*. In a marginal note accompanying the narrative, Purchas remarked that "this map of the author, with the tables of his journal and sayling, were somewhat troublesome and too costly to insert."

Baffin's had been perhaps the most magisterial and consequential of all the attempts thus far to find the Northwest Passage, but his findings almost immediately faded into obscurity. By the nineteenth century, most maps of the region no longer even included Baffin Bay. England had led the attempt to find routes through the Arctic to the east; and now, in the face of failure after failure, ice barrier after ice barrier, it turned its back on the pursuit. Noted author Jeannette Mirsky: "For two hundred years the wake made by the *Discovery* was lost; no other ship plowed the waters of Baffin Bay; and not until England had gone through her civil wars, settled and lost her colonies in America, and shipped Napoleon safely off to St. Helena did she turn north again."

That the rise of the second wave of Arctic exploration was concomitant with the fall of Napoleon I was not coincidental. In 1793, six years prior to Napoleon's seizure of power, the British naval fleet totaled approximately 500 vessels; by the time of the Battle of Trafalgar in 1805, that number had swelled to 950, with accompanying increases in personnel. With Napoleon's final exile in 1816, all those ships, officers, and men that had been specifically commissioned and recruited for the task of defeating his empire were suddenly without purpose.

Enter John Barrow: born in northwest England in 1764, he was appointed in 1803 as the second secretary—and thus principal bureaucrat—at the Admiralty, a position he occupied, outside of one brief interruption, for forty-one years. It was Barrow who proposed that there could be no better application of suddenly available personnel and materiel than toward exploration, particularly of the Arctic. It was a notion that received immediate and vigorous pushback: exploration cost money, and money was in short supply during a postwar era of austerity. But Barrow's pitch was artfully designed and skillfully delivered: it would be embarrassing—nay, humiliating—were Britain, which had been at the forefront of Arctic exploration two centuries previously, to now sit back as the glow of victory faded and allow rival nations to fill in the maps that Britons had outlined.

That argument struck home, earning Barrow the backing he sought and securing him funding in 1818 for a pair of synchronous expeditions, one of which sought to find the Northwest Passage and one of which probed for a path to the North Pole. Others followed, both under Barrow's aegis and independent of it, with varied degrees of success and privation. One, under the command of Sir John Ross, was forced to spend four winters trapped in the Northwest Passage's icy grip before being able to strike out in small boats and head far enough out into open water to be able to attract the attention of a passing whaler.

Virtually ever present in these efforts was Ross's nephew, James Clark Ross, who between 1818 and 1833 sailed on no fewer than six Arctic expeditions, two of them with his uncle, by the end of which he had surely spent far more time in the region than any European explorer. His reward was to be given command of his own expedition, to the other end of the world, mapping the coast of Antarctica aboard HMS *Erebus* and HMS *Terror*.

Erebus and *Terror* had been built as bomb ships—specialized naval vessels constructed and equipped to bombard land-based positions with mortars; stout and sturdy, their hulls built to withstand the recoil of launching powerful explosives in a high arc, they were theoretically ideal vessels for withstanding the pressures of polar ice.

Terror was the elder of the two; launched in 1813, it participated in the Battle of Baltimore, its bombs among those later immortalized by Francis Scott Key as "bursting in air." *Erebus*, built in 1826, endured an initially idle existence, other than a couple of years' service in the Mediterranean Sea, before sailing south with Ross.

Ross's Antarctic expedition, conducted between 1839 and 1843, was a resounding success, as testified to by the string of locations on the frozen continent—Ross Sea, Ross Island, the Ross Ice Shelf, among others—that today carry his name. Upon his return, he was asked to take the same two ships north on what was hoped to be a final, glorious voyage into the Northwest Passage, but he demurred; after four years in the Antarctic on top of four unplanned winters frozen into the Arctic and a handful of other Arctic expeditions, he had simply had enough.

Command of the expedition was ultimately bestowed upon Sir John Franklin, who had himself been a leader of a pair of Barrow's Arctic ventures—one by sea and one by land—before a spell as lieutenant governor of the British penal colony of Van Diemen's Land (now Tasmania). He did not, on the face of it, appear a tremendous fit: he was closing in on seventy years of age, and neither of his previous efforts had been particularly successful, with the overland venture ultimately something of a disaster. One of the members was shot dead by another, having been accused of killing four others and even feeding the flesh of three of them to some of the rest of the party; and, after Franklin waited too late in the year to turn the expedition back south, conditions became so dire that ten of the party died on the march home and the rest were reduced to consuming their footwear, earning Franklin the unwelcome sobriquet of the Man Who Ate His Boots.

Even so, Franklin was a genial and popular fellow, and the public mood was of great expectation as, on the morning of May 19, 1845, the two ships, with their combined complement of 129 men, one Newfoundland dog, one cat, and one monkey—the latter a gift to the officers of *Erebus* from Lady Jane Franklin—made their way out of the River Thames. They sailed up the North Sea to Stromness in Scotland and thence onward toward the Arctic. By late June, they had rounded Cape Farewell at the southern tip of Greenland and continued into Davis Strait. On July 26, they encountered the whaling vessels *Enterprise* and *Prince of Wales*; after a brief visit between the ships, the whalers continued on their way, leaving *Erebus* and *Terror* waiting for the ice to clear so they could sail into Lancaster Sound and the unknown.

In August 1848, with no word yet received from Franklin or his men, the Admiralty launched three search expeditions, none of which achieved success, denied either by searching in the wrong area or by impenetrable ice. Two years later, fully fifteen expeditions—some funded by the

Admiralty, others supported by private donors, all responding to the urgings of Lady Franklin—scoured the Passage for clues of Franklin's well-being or whereabouts.

The first such clue was uncovered on August 23, 1850, when Captain Erasmus Ommanney of the *Assistance* found stores and fragments of clothing on a beach on Devon Island, on the northern shore of Lancaster Sound, from where he spied a cairn on adjacent Beechey Island. The cairn, oddly, contained no note, nothing to confirm that Franklin had been there or to inform as to his subsequent whereabouts. But the discovery prompted several of the other vessels involved in the hunt to converge on Beechey, and one week after Ommanney saw the cairn, another search party made the first real breakthrough: signs of a winter camp on Beechey Island and, on the camp's edge, three graves, each marked with a wooden headstone.

The first of the three to have died was John Torrington, who, according to his tombstone, "departed this life January 1st A.D. 1846 on board of H.M. ship Terror, aged 20 years." He was followed just a few days later by John Hartnell, able seaman on *Erebus*, who "died January 4th 1846 aged 25 years." Hartnell's shipmate William Braine, part of the consignment of Royal Marines, died "April 3rd 1846, aged 32 years." There was no explanation of how or why they had died; nor did the rest of the campsite provide any answer at all. If anything, the findings only deepened the mystery. There were items of clothing, a few scraps of paper, pieces of rope, and fragments of sail. Bizarrely, there was a collection of six hundred food tins, stacked in a pile, emptied of their original contents and filled with gravel.

The searchers had found where Franklin had been but not where he had gone. The following year, a further five ships were dispatched; but after four of them were abandoned in the crushing ice with no further signs of success, enthusiasm for future search efforts began to pall.

Then, in April 1854, a party of four men, led by Scottish explorer John Rae and including interpreter and guide William Ouligbuck, were a few weeks into a journey to chart part of the coast of the Canadian Arctic Archipelago when they encountered two Inuit, named See-u-ti-chu and In-nook-poo-zhe-jook, the latter of whom was wearing a gold cap band. When Rae and Ouligbuck asked him where he had procured it, he replied that although he himself had not previously met any *kabloonas*, or white men, a number of them—at least thirty-five to forty—had starved to death west of a large river, and it had come from one of them. In-nook-poo-zhe-jook had not been to that place and could not tell them how far it was exactly, other than to suggest it was about a ten- to twelve-day journey. He either could not or would not locate it on a map, but he insisted that many dead bodies could be found there still. Rae bought the band from him and promised that he would pay for any similar trinkets that he or others might have.

Several weeks later, Rae and his party returned to the base from which they had set out, to find several Inuit families had pitched tents nearby and were waiting for them, with more objects to trade and more information to give. They told Rae and Ouligbuck that four years previously—that is, in 1850—several Inuit families hunting seals near the shore of King William Island had encountered about forty white men traveling southward over the ice and dragging a boat and sledges. None of the men could speak Inuktitut, but through sign language they indicated that their ships had been crushed in the ice and that they were heading to where they expected to find deer to shoot. The men appeared thin, and they traded for a small seal, or a piece of seal meat, from the Inuit.

Later that same season, Rae reported his interlocutors as telling him that the "bodies of some thirty persons and some Graves were discovered on the continent, and five dead bodies on an island near it." Some of the bodies, he continued, "had been buried (probably those of the first victims of famine); some were in a tent or tents; others under the boat, which had been turned over to form a shelter, and several lay scattered about in different directions." From its description—a telescope strapped over its shoulders and a double-barreled shotgun underneath it—Rae inferred that one of the bodies was of an officer.

Their testimony was given extra credibility by the items those who gave it had in their possession and sold to Rae: a brass compass, a chronometer case, a surgeon's knife, and forks and spoons engraved with the initials of some of the officers onboard *Erebus* and *Terror*, including Harry Goodsir, assistant surgeon on *Erebus*, and Francis Crozier, captain of the *Terror* and the expedition's second-in-command. Any lingering uncertainty as to exactly which group of men the objects belonged was disabused by one item in particular, a silver plate engraved with the name Sir John Franklin.

There could no longer be any doubt. The Inuit had seen the survivors of Franklin's expedition; but if the physical evidence reinforced the veracity of their accounts, it also made them all the more chilling, particularly given one detail that drowned out all the others: "From the mutilated state of many of the bodies," wrote Rae, "and the contents of the kettles, it is evident that our wretched countrymen had been driven to the last resource—cannibalism—as a means of prolonging existence."

Rae contained all these details in a report to the Admiralty, which swiftly made its way to the pages of the *London Times*, where it detonated with a concussive impact that reverberated throughout the corridors of British society. The suggestion that Franklin's men had resorted to cannibalism set public opinion ablaze, fanned by nobody more so than Charles Dickens, friend of Lady Franklin and the country's foremost author, who used his own magazine, *Household Words*, to fulminate in righteous anger. Dickens, seeking solace in a mixture of equal parts patriotism and racism, excoriated Rae's reports from his Inuit informants as "the vague babble of savages." He insisted that if mutilated corpses had indeed been found, then the "covetous, treacherous, and cruel" savages who claimed to have found them were among the likelier mutilators. Certainly, he insisted, the officers and men of the *Erebus* and *Terror* could not have committed such heinous acts; "the noble conduct and example of such men, and of their own great leader himself, … belies it, and outweighs by the weight of the whole universe the chatter of a gross handful of uncivilised people."

Even so, Dickens conceded that the physical evidence led to an inevitable conclusion that the expedition's members were dead. Lady Franklin was unsatiated, but the Admiralty made it clear that it would not be moved to send yet another search party; apart from the growing pile of corpses and shipwrecks accrued in the search for the Northwest Passage and now for Franklin, its resources were once more tied up in war, this time in Crimea. She was, however, able to persuade Captain Francis Leopold McClintock, veteran of two searches for her husband, to make one more, with an all-volunteer crew of twenty-five aboard the steam yacht *Fox*. She acknowledged that Sir John, seventy years old when he had departed a decade earlier, was surely dead; her fight now was for his legacy, her desire to disprove Inuit tales of cannibalism and establish that he had pieced together the segments of the Northwest Passage.

McClintock was not able to establish the definitive proof she sought for either of those goals, but he was able to provide confirmation of a different kind. A series of sledge parties from the *Fox*—led by McClintock; his second-in-command, William Hobson; and the ship's master, Allen Young—found enough evidence on King William Island to ascertain that it was indeed there that at least the majority of the *Erebus* and *Terror* crew met their fate. Encounters with Inuit on the island elicited descriptions of a three-masted ship "crushed by the ice out in the sea to the west of King William's Land" and later of a second ship, "which was forced on shore by the ice." There were stories of the body of a "very large man" with "long teeth," found onboard one of the ships; of white men heading south to a "large river" and of their bones being found the following winter; of men who "fell down and died as they walked along."

Then, on May 6, 1859, Hobson's sledging team found a cairn containing a cylinder. In the cylinder was a printed form, one of many sent with the *Erebus* and *Terror* for recording their position, filled in by Lieutenant Graham Gore of the *Erebus*:

> 28 May 1847 H.M.S.hips Erebus and Terror Wintered in the Ice in Lat. 70°5'N Long. 98°23'W. Having wintered in 1846–7 at Beechey Island in Lat. 74°43'28 N. Long. 91°39'15 W. after having ascended Wellington Channel to Lat. 77° and returned by the °West side of Cornwallis Island. Sir John Franklin commanding the Expedition. All well. Party consisting of 2 Officers and 6 Men left the ships on Monday 24th May 1847—Gm. Gore, Lieut., Chas. F. Des Voeux, Mate.

Gore had made a mistake: given the dates on the Beechey Island graves, it was clear the expedition had wintered there a year earlier than stated in the note. Such a concern was secondary, however, to the information detailed in the writing around the form's edges, added eleven months after the original note and revealing that the situation was now far from "all well":

> 25th April 1848 HMShips Terror and Erebus were deserted on the 22nd April 5 leagues NNW of this having been beset since 12th Sept 1846. The officers and crews consisting of 105 souls under the command of Captain F.R.M. Crozier landed here.... Sir John Franklin died on the 11th of June 1847 and the total loss by deaths in the Expedition has been to this date 9 officers and 15 men.— James Fitzjames Captain HMS Erebus F.R.M. Crozier Captain & Senior Offr And start on tomorrow 26th for Backs Fish River.

Elsewhere on the island, McClintock found a skeleton of a man, face down and partially dressed, later identified as likely being a steward from *Terror*; nearby on the ground, a pocketbook contained a series of drawings and writings, the latter of which McClintock presumed to have been in German but were in fact in English, yet often spelled both phonetically and backward.

Farther up the coast, a large boat sat entombed in drifting snow, close to a sledge on which it had almost certainly been dragged. The boat contained an abundance of supplies, some surprising: blankets, boots, and clothing but also books, plates, and engraved forks, knives, and spoons. In the boat also were a pair of rifles and two more skeletons—one of a seemingly strong and fit middle-aged man, one of someone younger and slighter.

Once again, the discoveries raised more questions than answers. What had happened between the message that all was well and the death, two weeks later, of Franklin? Why, apart from the so-called Victory Point note, were there no other descriptions of the expedition's whereabouts and plans? Why was the bow of the boat containing the two skeletons pointed north, toward the abandoned ships, rather than south, the direction in which the men were traveling? And where were the ships?

Nonetheless, the underlying mystery was solved. John Franklin was dead, his men with him. The search for the Northwest Passage had consumed vast amounts of money, time, materiel, and lives. The answer it had yielded—that a passage, or rather passages, existed, but seemingly at no point free of ice in totality—was emotionally unsatisfying, and all the more so given the effort required to provide it. A pursuit that had been given impetus by a treaty between Portugal and Spain had reached a deadly conclusion four centuries later, a quest fueled by dreams and aspirations wrecked on the icy shores of reality.

The search for the Northwest Passage was over.

———⟡———

A century and three decades after McClintock confirmed the outline, if not the details, of Franklin's fate, Dr. James Hansen took his seat in a committee room of the United States Senate. It was a stifling June day in Washington, DC, with the mercury nudging close to 100°F; and Hansen, head of the National Aeronautics and Space Administration's Goddard Institute for Space Studies, testified that the heat wave was not an anomaly and that human activity was responsible for increasing temperatures across the globe.

"The earth is warmer in 1988 than at any time in the history of instrumental measurements," Hansen told the senators, adding that there was "only a 1 percent chance of an accidental warming of this magnitude.... The greenhouse effect has been detected, and it is changing our climate now."

The echoes from Hansen's alarm reverberated widely. The following day, the *New York Times* reported that "global warming has begun." But while

Hansen's testimony was a key early moment in alerting the United States to the realities of global warming, and while 1988 was a pivotal year in terms of mobilizing the international community to address the threat it presented—the United Nations General Assembly endorsed the establishment of the Intergovernmental Panel on Climate Change that December—the warning bells had been ringing for much longer.

The scientific fundamentals were laid down as far back as 1824 by French physicist Joseph Fourier, who wondered why, given that Earth was constantly being bombarded by solar radiation, the planet did not become progressively hotter until it was a baking and barren rock. The answer, he deduced, was that much of that radiation was reflected out into space, but that gases in Earth's atmosphere were trapping some of the heat—a phenomenon that came to be known as the "greenhouse effect." Thirty-five years later, British scientist John Tyndall identified the gases most likely to be responsible for the planet's warming blanket: methane, water vapor, and carbon dioxide (CO_2). It was Swedish scientist Svante Arrhenius who in 1896 calculated that a small increase in CO_2 concentrations might warm the atmosphere sufficiently to allow the formation of greater quantities of water vapor—which in turn could lead to further warming. Such increases, Arrhenius noted, were not entirely hypothetical: the Industrial Revolution had been made possible by, and resulted in, the combustion of large amounts of fossil fuels such as coal—a process that had released the carbon those fuels contained into the atmosphere, where it had combined with atmospheric oxygen to create CO_2. If the levels of atmospheric CO_2 were to double, he calculated, Earth could see warming of up to 8°F or 9°F; fortunately, he estimated, such an eventuality would be unlikely for another two thousand years or more.

Barely four decades later, English steam engineer and amateur climatologist Guy Stewart Callendar published evidence that not only had global land temperatures increased over the previous fifty years, but this increase correlated with higher concentrations of CO_2 in the atmosphere. In the 1950s, Charles Keeling calculated that the level of atmospheric CO_2 was approximately 310 parts per million (ppm)—that is, for every million molecules of atmospheric gas, 310 were CO_2—and began a series of daily measurements from atop Mauna Loa in Hawaii, which have shown that atmospheric CO_2 levels continue to trend upward.

Prior to the Industrial Revolution, the concentration of CO_2 in the atmosphere was approximately 285 ppm; by the time Hansen addressed the Senate, it had climbed to 350 ppm, and as I write this in mid-2024, it is roughly

426 ppm. During that time, and contra Arrhenius's belief that major warming would unfold on a timescale of centuries or even millennia, average global temperatures have already climbed by approximately 2°F since he made his calculations, with two-thirds of that warming taking place since 1975. Such increases, however, are not uniform; while the world has been warming, the Arctic has been warming most of all, roughly twice as much as the global average. In 2020, temperature across the region was approximately 4°F higher than the average between the years 1981 and 2010.

There are several reasons why the Arctic should be especially prone to warming. The "weather layer" of the atmosphere known as the troposphere is thinner above the poles than at the equator and so requires less energy to create a given amount of warming. Additionally, in the humid tropics, a sizable proportion of the Sun's energy is expended on evaporation; in the dry air of the Arctic, that energy leads directly to heating. And the Arctic contains a number of feedback mechanisms that exacerbate warming, primarily in the form of what is known as albedo: snow and ice cover in the polar regions reflects sunlight, meaning that those areas have a high albedo, but as that snow and ice melts, it exposes darker rock and water underneath, which absorbs more heat, prompting more heating and further melt.

Since 2007, the National Oceanic and Atmospheric Administration (NOAA) has been documenting observed year-on-year changes in the Arctic environment in an annual Arctic Report Card. Some of the more recent and notable documented impacts of increased Arctic temperatures: record low June snow cover in Siberia; extensive wildfires in northern Russia; "browning" of the tundra in North America; rainfall on the Greenland ice sheet; thawing permafrost that damages infrastructure and, in turn, releases additional stores of carbon into the atmosphere. But arguably the most visible, dramatic, and easy-to-comprehend change is in the extent of Arctic sea ice.

From mid-September until mid- to late March, Arctic sea ice steadily expands in area before decreasing with the return of warmer weather and longer days until it reaches its minimum extent and the cycle begins anew. Since 1979, microwave sensors on satellites have provided uninterrupted imagery of Arctic sea ice, and from the earliest years of analysis, there has been evidence that the extent of that sea ice was in decline. Initially, however, that decline was halting; for the first dozen or so years of observation, low summer ice coverage would be followed by a return to average, or near-average, conditions in the winter. That changed in 2002, when the September sea ice minimum fell to a record low of 2.3 million square miles, more than 400,000 square miles below the 1979–2000 average and almost 66,000 square miles less than the previous record low. The next few years saw a mild rebound, another fall in 2005, and then, in 2007, a collapse.

That year, a confluence of atmospheric conditions laid waste to an ice cap that had not recovered strongly from the previous summer. The result was a September minimum sea ice extent of 1.65 million square miles, 23 percent lower than the previous low and 39 percent lower than the average during the satellite era. "Huge chunks of ice were missing," Mark Serreze, director of the National Snow and Ice Data Center, later wrote. He described viewing the collapse with "a sense of morbid fascination. . . . I had never seen anything like this before." In an interview with the *Washington Post* thirteen years later, he described that year as "the beginning of the 'new Arctic.' Since 2007 it [has] been one piece of bad news after another."

As epochal as 2007 was, it would soon be surpassed. The year 2012 usurped its position as having the lowest sea ice minimum on record, and 2020 slotted into second place, while 2016 and 2019 would join 2007 in a statistical tie. As of July 2024, the seventeen lowest Arctic sea ice years were the seventeen most recent Arctic sea ice years.

Serreze has described Arctic sea ice as being in a death spiral as more Arctic warming begets further Arctic melting, which facilitates additional warming and yet more melting, the process potentially continuing absent correction until the cycle may be unstoppable.

(Although similar, "area" and "extent" are not quite the same. Area is the total amount of surface covered by sea ice; extent is the entire space within the outer boundary of sea ice coverage, including areas of open water. In the words of the National Snow and Ice Data Center, "A simplified way to think of extent versus area is to imagine a slice of Swiss cheese. Extent would be a measure of the edges of the slice of cheese and all of the space inside it. Area would be the measure of where there is cheese only, not including the holes.")

While much focus is on the declines in Arctic sea ice area or extent, of arguably greater import is the even more dramatic decline in sea ice *volume*—and, at the same time, in its age.

Look at a graphic of Arctic sea ice extent over the course of a year and two things become immediately evident: there is, unsurprisingly, far more ice in winter than in summer; but there is some degree of sea ice coverage year-round. Some of the ice that forms in winter, in other words, persists through the following summer and into the following fall, and some of that ice survives yet further, into a third year or even beyond. And the longer sea ice lasts, the more likely it is to continue to last because with each successive winter, it becomes progressively thicker: while first-year ice may be just a few feet thick, so-called multiyear ice can extend to a depth as great as fifteen feet. But increasing temperatures not only prevent ever-larger amounts of fresh ice from surviving into a second year; they also chip away at the older ice, reducing overall sea ice cover not only in extent but also in thickness and thus volume. According to data compiled by the Polar Science Center at the University of Washington, average Arctic sea ice volume in November has fallen from 4,828 cubic miles (mi^3) in 1979 to 3,781 mi^3 in 2001 and 2,254 mi^3 in 2021. According to the National Snow and Ice Data Center, the amount of multiyear ice coverage in the Arctic declined from 61 percent in 1984 to just 34 percent in 2018, and just 2 percent was composed of the thickest, oldest ice that had persisted for five years or longer—raising the prospect, said Serreze, of a future in which multiyear ice could effectively cease to exist and "sea ice will be but a seasonal feature of the Arctic Ocean."

That would matter for multiple reasons. Species such as polar bears that live and hunt on the surface of the sea ice would be most immediately and obviously at risk of depletion in, or extirpation from, those regions where their habitat was but a temporary feature; so, too, the ice-dependent species on which they prey. Thinning or absent ice, and any consequent alteration in the finely balanced timeline of ice melt and growth, could have comprehensive and complex consequences for the entire Arctic marine ecosystem to an extent that scientists are only beginning to unravel. For the people who live and subsist in the Arctic, and for whom sea ice is not just integral but fundamental to their history, culture, and survival, the changes would be profound, threatening the existence of a way of life that has thrived for millennia. Although the evidence remains inconclusive, declining Arctic sea ice may also have significant impacts on climatic conditions in lower latitudes as warmer temperatures disrupt oceanic and atmospheric circulation patterns.

And yet, where many see disruption and challenges, others sense opportunity, and not just in the opening of possible trade routes. In 2007, a Russian submarine planted a titanium flag on the seabed at the North Pole, and in October 2019, Russia's ministry of defense proclaimed that it had collected enough evidence to support a territorial claim to much of the Arctic Ocean. Canada has exerted a counterclaim, while Denmark has laid down its own metaphorical marker, based on the fact that a large area up to and beyond the North Pole is connected to the continental shelf of Greenland.

Some observers worry about the impact of increased traffic and development on wildlife and the environment, in the form of increased noise and development and the potential for accidents, oil spills, and pollution. Others envisage a scenario in which regional development and economic growth empower and enrich the peoples of the North. Some see competition, the emergence of a truly cold war as great powers jostle over issues of access and territory. The United States military reportedly frets about an "icebreaker gap"; defense think tanks compare growing tensions in the Arctic to those in the South China Sea; and the first Donald Trump administration clumsily declared a desire to purchase Greenland and responded in a huff when the notion was airily dismissed by Denmark. Yet there is a case to be made that such conjecture is overwrought and overblown, particularly given the existence of established treaties that promote a history of cooperation within the Arctic and the fact that Arctic nations are merely conducting due diligence in anticipation of the region ultimately becoming at least seasonally ice-free.

What is clear, however, is that absent steps to address climate change, the Arctic will undergo significant change in the coming decades. Indeed, it is already doing so, to an extent that the Arctic of today would in many ways be barely recognizable, both to those who first searched for passages through its frozen waterways and to those whose presence preceded them by centuries.

Northwest

CHAPTER 1

A Highway to Everything and Everywhere

The Canadian Arctic Archipelago stretches approximately 1,500 miles from east to west and 1,200 miles from the Canadian mainland in the south to Cape Columbia, which at just 478 miles from the North Pole is the most northerly point of land anywhere outside of Greenland. The archipelago covers approximately 550,000 square miles and comprises 36,563 islands, 3 of which—Baffin, Victoria, and Ellesmere—are respectively the fifth-, eighth-, and tenth-largest islands in the world. While much of the central and western archipelago is low-lying, the major islands to the east are mostly mountainous; Barbeau Peak on Ellesmere is, at 8,583 feet, the highest point in eastern North America. Those same eastern islands contain, in their higher reaches, 75 percent of the glacier ice in Canada and one-third of the volume of land ice on the planet outside of Greenland and Antarctica.

One definition of what constitutes the southern limit of the Arctic is a line along which the mean temperature of the warmest month is 50°F (10°C); denizens of the Canadian Arctic Archipelago can mostly only dream of such warmth. In the islands' most northerly reaches, the mercury may nudge above freezing for only a month or two, and the average year-round temperature can be as low as −4°F (−20°C), with extreme lows below −50°F (−46°C). Even the lower latitudes of Baffin Island, one of the more southerly areas of the archipelago, may experience average annual temperatures of just 22°F (−6°C). This is not merely the Arctic but the High Arctic, a region characterized by prolonged low temperatures and extremely low precipitation—less than three inches per year on parts of Ellesmere Island. The challenging conditions are reflected in the inventory of flora and fauna: a mere twenty species of land mammals live on the archipelago, and while that list includes the mighty musk ox, it also features a miniaturized subspecies of caribou, the females of which weigh on average only 130 pounds. Plant life is dominated by dwarf shrubs, sedges, grasses, 325 species of mosses, and up to 600 species of lichens. There are no trees.

It is in the water—the bays and channels that surround the islands, that separate them from each other and the mainland, that constitute the North-west Passage and its tributaries and dead ends—that the region truly comes alive. The cold waters are rich in nutrients that support marine phytoplank-ton—single-celled photosynthesizing organisms that form the base of the marine food web and that are preyed upon by microscopic animals called zooplankton: miniature crustaceans known as amphipods and copepods and floating mollusks called pteropods.

All of these—phytoplankton, amphipods, pteropods, copepods, and more—support an ecosystem that, in the archipelago region, includes sixty-eight exclusively marine fish species such as Arctic cod and a few anadromous species (which spend parts of their lives in both fresh water and salt water), most notably Arctic char, as well as marine mammals including ringed and bearded seals, walruses, belugas, narwhals, bowhead whales, and polar bears. That a bear should be included in a list of marine mam-mals might on the face of it seem surprising, but polar bears are officially classified as such—not because their relatively long necks and massive, paddle-like paws make them especially able in the water, nor because they have on occasion been tracked swimming tens or, exceptionally, hundreds of miles. Polar bears are marine mammals because they depend on prey from the ocean. They very rarely, however, hunt that prey—primarily ringed and, to a lesser extent, bearded seals and occasionally belugas or even wal-ruses—*in* the ocean, where they would be at an enormous disadvantage; rather, the great majority of the time, they hunt *on* the ocean, specifically on the frozen surface known as sea ice. It is on sea ice that ringed seals den and pup; and it is in sea ice that they use the long claws on their flippers to carve breathing holes, next to which polar bears may lurk.

But sea ice is more than a frozen surface on which polar bears perambulate and prey. Sea ice is not merely a characteristic of the northern polar realm; it is the defining feature of Arctic seas, the attribute that allows the region's marine ecosystems to not only survive but thrive. It is not just the engine that drives Arctic marine life; it is the spark that brings that engine to life and the fuel that keeps it running.

As sea ice forms, it traps multitudes of microscopic organisms within it, including unicellular phytoplankton called diatoms. These diatoms have evolved specifically to survive in ice floes; despite being plants that need photosynthesis to survive, they are able to move, and so they do—but not toward the top of the ice, where sunlight may be more likely to reach, but to its bottom, where they bide their time and wait for spring.

Although sea ice is more saline than freshwater ice, it is fresher than the water from which it formed because some salt is extruded during its formation. (The older the ice, the fresher it is.) With the return of spring, sunlight penetrates the ice and warms the water below, melting the floes from the bottom and creating a layer of relatively fresh water, complete with diatoms, which—because fresh water is lighter than salt water—floats at the surface. The Sun's rays bathe it with warmth and light, the diatoms bloom, and the Arctic springs to life. The diatoms are eaten by small zooplankton, small zooplankton are eaten by larger zooplankton, and so on, through cod and char and seals and seabirds and whales and bears, the melting ice and the returning Sun setting in motion an explosion of life, of feeding and breeding and birthing that is strong enough to sustain the Arctic marine ecosystem through summer and into early fall. (With some topping up: while the spring diatom bloom is by far the most significant, nutrients continue to be brought into the photic zone—the layer of water that sunlight can reach—via deep-lying waters from the Pacific Ocean, prompted by storms, tidal action, or natural upwelling over steep-sloping seabed.)

The notion of sea ice as life force would not have been foremost in the minds of the first European explorers who vainly probed and prodded its resilient blockade in their search for a Northwest Passage. To them, it was a merciless foe, ready not just to impede but to entrap and destroy, a malevolence of almost unimaginable power that prevented progress and threatened so much more.

But to those whose presence in the region preceded such interlopers, sea ice was and remains an integral component of life. Indeed, as noted by Claudio Aporta of Dalhousie University, who has extensively studied Inuit relationships with sea ice in the Canadian Arctic, the difference in Indigenous and Western approaches is summed up in the terminology they use. For the latter, the channels through the Canadian Arctic Archipelago—whether frozen or thawed—are a passage, a place through which to travel between oceans. It was thus during the age of exploration, and it remains so now: the Arctic, he wrote, "is still a transitional place where scientists conduct their fieldwork, companies attempt to find minerals and other resources, and tourists go to visit." In contrast, the geographic space known to us as the Northwest Passage, and the coast and seas around it, have been perceived and occupied in a completely different way by Inuit. For them, sea ice has never been an unwanted visitor or a blocking presence. It is, in fact, the opposite: for Inuit, the sea ice is both a social surface and a place to call home. The idea of the Arctic seas as a body of water connecting two oceans was not part of the Inuit view of the world, but they regularly travelled across long expanses of land and sea ice, through significant parts of what we now refer to as the Northwest Passage, and across the Canadian Arctic Archipelago. The open and frozen sea was simply part of a territory where life took place, as opposed to the European perception of the passage as a discrete geographic entity.

Far from seeing it as a passage, Inuit refer to the Canadian Arctic—not just the islands of the archipelago but also the northernmost reaches of the mainland, from Arviat on western Hudson Bay to Grise Fiord on Ellesmere Island, from Tuktoyaktuk and Aflavik in the west to Qikiqtarjuaq and Makkovik in the east—as Inuit Nunangat. The phrase translates approximately as "our homeland" but encapsulates far more than the land itself: "When defining our 'land,' Inuit do not distinguish between the ground upon which our communities are built and the sea ice upon which we travel, hunt, and build igloos as temporary camps. Land is anywhere our feet, dog teams, or snowmobiles can take us."

In the foreword to a 2014 study of Inuit reflections on sea ice, Duane Ningaqsiq Smith of the Inuit Circumpolar Council reflected:

> Sea ice has always been part of my life. I spent much time as a child and as a teenager travelling to the coast and across the ice. I never questioned why it was there. It just was. But I did know it was special. It brought me, and my grandfather who almost always accompanied me, to places that I found exciting. It brought me to places that sustained us. We secured our livelihood on the ice. We hunted and we transported goods. We visited hunting camps of relatives and friends. We studied the ice and predicted weather and animal migration patterns. . . . It connects us to places, to people and to our environment. It is our highway to everything and to everywhere.

This is not simply a case of a people adapting to the challenges of a hostile environment. Arctic waters, and particularly frozen ones, are the central element of Inuit life and culture. With few exceptions, Inuit settlements are built on the coast or within easy reach of summer sea and winter ice. Sea ice is the primary means of transport from one such settlement to another: indeed, the word *tusartuut*—which Inuit have traditionally used to describe the time when landfast ice becomes thick enough to permit travel—means "news time" or "when one is able to hear from other camps." (Landfast ice is sea ice that is anchored to the coastline or shallow seabed. Because it does not move with wind or currents, it is inherently more stable than pack ice and thus better suited to travel.) Sea ice—and especially the areas where sea ice meets water, at the ice's edge or along cracks or around open areas in the ice known as polynyas—are where Inuit hunt seals, walruses, or whales or fish for Arctic cod. Without sea ice, Inuit life as we know it, and as it has long been, not only would not be the same; it simply would not be.

And yet, in a casual illustration of the degree to which, to the outside world, sea ice is little more than an ephemeral oddity at best and a transient obstacle at worst, Aporta noted that in 2009, Google Earth removed its sea ice coverage "in order to provide a layer of undersea bathymetry." As Aporta observed, this most unique of environments, which supports life throughout much of the Arctic and is the historical and continuing basis of an entire civilization, was deemed by those elsewhere to be of less interest and import than what lies beneath.

The first people to arrive in the Canadian Arctic did so approximately four thousand years ago, moving east from Siberia and Alaska, spreading out across the Arctic mainland and the islands of the archipelago and into Labrador and Greenland. They lived in small settlements, generally on or near the coast, characterized by semi-subterranean houses. They hunted seals and walruses in the frozen waters of the region and caribou on the tundra in the lower Arctic, their range expanding and contracting in accordance with shifts in the region's climate. For a period of about a thousand years, from approximately 550 BCE to a little after 400 CE, they withdrew from the High Arctic extremities such as Ellesmere Island entirely, in response to a deepening cold across the Northern Hemisphere. A shift from that cold to a steady warming allowed them to flourish, but it would also bring about their demise.

Around 1000 CE, on the west coast of Alaska, a new culture was emerging that would supplant the one that at that time occupied the Central Arctic region. Like that preexisting culture—dubbed the Dorset people by archaeologists, after Cape Dorset in the Canadian Arctic Archipelago, where their artifacts were first found in 1925—this emerging culture hunted marine mammals, but they had developed additional skills and technology that made them especially adept at finding and hunting seals along the ice pack. They used dog teams so that they could travel greater distances across the ice. And they learned to hunt bowhead whales, developing skin boats called *umiat* (a single boat is an *umiak* or *umiaq*), in which to travel from the ice edge into the open water in pursuit of them, and harpoons with which to strike them. Their success in bowhead hunting enabled them to live in what were, by Arctic standards, relatively large communities, given that a solitary forty-ton bowhead could feed five families for a year while also providing oil for heating and lighting and bones and baleen for buttons, tools, and utensils. This culture's emergence coincided with another warm period, which resulted in sea ice retreating farther north, freezing later, and thawing earlier, a development that in turn enabled bowheads to expand their range into the Central Arctic waters of Canada. According to this theory, their pursuers—dubbed the Thule people—followed their prey east from Alaska into the islands of the Canadian Arctic Archipelago, followed by a second wave who inhabited more southern Arctic waters. More recently, there has been some pushback against this explanation, arguing that it essentially paints the Thule as passive beneficiaries of a warming climate and that conventional wisdom does not quite match up with available evidence. This revised theory postulates that the bulk of the Thule migration took place two centuries later, when conditions would not have been as conducive to bowheads filling the Central Arctic, and that a primary motivation was the search for iron, a rare commodity in the Arctic.

The conditions that enabled expansion by the Thule also worked against the Dorset. As Renée Fossett of the University of Manitoba has observed, shorter, milder winters and a decrease in sea ice would have greatly impinged on the Dorset people's ability to hunt seals, while a changing climate would have also altered caribou migration patterns and made the herds more dispersed. For a culture that had essentially lived much the same life for close to two millennia, such changes may well have come too swiftly for them to adapt.

It was not long after their arrival, however, that the Thule were themselves forced to respond to a changing environment. By the 1300s, conditions were once more starting to cool; that, and the fact that bowheads now rarely ventured into the shallow channels of the central Northwest Passage, meant that they diversified their hunting targets and techniques, in many ways replicating the Dorset by focusing, in parts of their range, on sealing, fishing, and hunting caribou. Over the next several centuries, they retrenched from and returned to parts of their range as conditions required; they abandoned the Queen Elizabeth Islands in the far north, never voluntarily to return. Their settlements became smaller and more mobile, and they adopted a more nomadic way of life, making them more readily able to move to where resources were available. By the 1500s, the Thule had effectively become modern Inuit, and the Dorset had to all intents and purposes long since disappeared. Inuit traditions tell of a people, whom the Inuit call Tunit, who lived in the region when their ancestors arrived; according to these traditions, Tunit and Inuit "hunted in company and were good friends" for a period after the Thule's arrival. Indeed, some archaeological sites show signs of being occupied simultaneously by Dorset and Thule, although others were occupied by one and not the other, suggesting, according to Renée Fossett, "that the two groups had different ideas about what constituted a good residential site."

Whether through assimilation, competition, violence, or some combination thereof, the Thule appearance led by the 1200s to the Dorset disappearance; it is possible also that the spread of Thule into Greenland a couple of centuries later may have hastened the end of the Norse colonies there.

Perhaps a hundred or so years after the Greenland colonies disappeared, Inuit on an island just off southeast Baffin Island watched as a ship appeared offshore. It was the *Gabriel*, commanded by Martin Frobisher, the ship that just a few weeks earlier had been waved farewell by Queen Elizabeth I and that now was probing the coastline of the Canadian Arctic in search of the Northwest Passage. The locals viewed the strangers warily as they came ashore, and vice versa; initially, however, their interactions appeared to be positive. To ensure mutual safety, one Inuk went aboard the *Gabriel* while two sailors stayed ashore with the villagers, and both sides conducted some trade. But when five men from the *Gabriel* rowed to the village with their guest, none of them returned; in retaliation, the *Gabriel*'s crew kidnapped another villager and, with the Arctic summer rapidly ending, departed with him for England, where he fell ill and died. Several centuries later, nearby Inuit recounted a tale that the five crewmen, having been captured, overwintered at the village and later built a boat and set off into open water, never to be seen again.

Similar scenes unfolded when Frobisher returned the following year: a battle that left a half dozen Inuit captured; another three Inuit—including a mother and young baby—kidnapped and taken to England. A few years later, John Davis's encounters with Inuit followed a similar, albeit less extreme, pattern: spells of fruitful interaction, conducted against a background of mutual wariness, interrupted by occasional, seemingly sudden, skirmishes.

For both sides, these encounters were surely infused with misunderstanding and suspicion, not least as they took place somewhat out of the blue, upon the sudden appearance of the visitors' ships and over a few days or at most weeks. Similar scenarios played out elsewhere in the region over the coming years and decades: only eight of the thirteen mutineers who set Henry Hudson adrift were still alive by the time the expedition ship reached England, as five were killed while ashore. Over time, interactions became more common and the two sides more familiar with each other as whalers and fishers appeared with increasing frequency off the east coast of Baffin Island and fur traders established posts in the Hudson Bay region, trading with Inuit as well as Athabascan and Chipewyan Indians. But not until the nineteenth century did the outsiders find a way past Baffin Island and penetrate the waters of the Canadian Arctic Archipelago; when they did so, their visits were not brief. Their ships frequently became lodged in the winter ice, the crews forced to endure what to them were unfamiliar and hostile surroundings as they waited for a spring thaw to release them. For some, however, that thaw never arrived, and, without the most basic of survival skills or adequate clothing and shelter, they met their end in a world they had hoped to traverse, conquer, and claim, but in which they never truly belonged.

CHAPTER 2

The Place That Never Thaws

One hundred and seventy-four years after *Erebus* and *Terror* sailed into what would become their icy tombs, I stood on the deck of a very different ship in a very different time as it idled along the coast of Devon Island on its way through Lancaster Sound. That ship, the *Ocean Endeavour*, had originally been built as a Baltic Sea ferry and was now operating as a passenger vessel, transporting its paying guests with consummate ease through the same Northwest Passage that had bedeviled Sir John Franklin and his predecessors, with nary a piece of sea ice to be seen along the route. I felt the breeze on my face and watched the afternoon light radiate off a small iceberg as the ship approached Dundas Harbour, one of the last—or, if traveling westward, first—bays guarding the Passage's entrance and exit.

In Inuktitut, this land is known as Tallurutit, or "Woman's Chin with Tattoos," a reference to the appearance, from a distance, of the crevasses that streak up and down the island's cliffs. Sir William Edward Parry dubbed the island Devon, after the English county, during an 1819 voyage that took him three-quarters of the way across the Canadian Arctic Archipelago. In a warmer climate, Dorset and Thule peoples had made settlements here, but the island had been long abandoned by the time Parry appeared, and it would not be permanently inhabited again for another century and change. In August 1924, the Canadian government ship *Arctic* deposited three Royal Canadian Mounted Police officers, with fuel and provisions, to establish an outpost in Dundas Harbour:

> By official accounts, the buildings were "erected without mishap and the stores were placed in a storehouse situated about a quarter of a mile from the living quarters," no doubt a precaution against marauding polar bears. They passed a quiet winter, seeing no one and scarcely any wildlife. They travelled only very little, due to the "inhospitable nature of the interior" of Devon Island and the rugged ice of the frozen sea. For the first

three years of their existence at the post, the men did not even have radio contact with the south. They waited for the annual visit of a government ship to receive the year's news.

In 1926, one of the three men, Constable Victor Maisonneuve, committed suicide. The following summer, another constable, William Stephens, accidentally shot himself while hunting walruses. They are buried, along with the daughter of one of the Inuit families who helped keep them alive, in a small cemetery, Canada's most northerly, just up a slope behind the buildings—which, although long abandoned, still stand and are tended to, as are the graves, by the RCMP on an annual basis.

I stood alone, lost in thought as our ship approached the harbor's entrance, until Marc Hébert appeared silently beside me. The previous year, Marc and I had been on the shores of Hudson Bay—me as a journalist and broadcaster making one of my periodic pilgrimages to Churchill, Manitoba, the "polar bear capital of the world," he as a guide and driver of the unique giant-wheeled, school bus–size "buggies" that carry tourists across the tundra in search of polar bears that are waiting for the bay to freeze up so they can spend the long winter patrolling its icy surface in search of seals. Now, he was performing a similar function as inflatable driver, expedition team member, and onboard expert for Adventure Canada, which had chartered the *Ocean Endeavour*; I, at Adventure Canada's invitation, was fulfilling a long-standing dream to travel the Northwest Passage and follow somewhat in the footsteps of Franklin.

"Man, I want to find a bear," he confided in his soft French-Canadian lilt. He scanned the hills and beaches with his binoculars before slipping away as quietly as he had arrived. Almost immediately, Pierre Richard, the ship's naturalist, appeared in his place with his spotter scope.

"There's a bear on the point," he announced just minutes later as he peered through the scope's eyepiece in its direction.

I scanned the area, struggling but failing to see what he was looking at, even as the returning Marc pointed it out directly to me and the deck area filled up with passengers who had been alerted to the bear's presence. Suddenly, there it was: what appeared to be a large, healthy male polar bear feasting on the carcass of a beluga whale on the beach.

No sooner had that one bear caught our collective attention than Marc spotted another, on the hill across the bay from the first, and then almost immediately two more, seemingly a mother and a large, approximately two-year-old, cub. And meanwhile, the sky overhead and the water up ahead were filled with gulls, belugas were surfacing near the coast, and harp seals chested their way en masse through the harbor. We had stumbled across a feeding frenzy, an abundance of Arctic cod fueling a polar smorgasbord.

The plan had been to step ashore, visit the RCMP hut and the graves, and take a short hike to a nearby Thule settlement. The presence of so many polar bears ashore naturally put an end to that idea but saw it replaced with another: we would put boats in the water and take a tentative close-up look at the wildlife action unfolding in front of us.

It was an extraordinarily beautiful day, the Sun shining and the water still, as our boat, with Marc driving, carved a gentle wake on its way into the bay. We coasted as we watched the mother and cub clamber up the side of the hill and out of sight and as bobs of seals passed by the boat, oblivious to our presence and focused only on the explosion of Arctic cod beneath the waves.

The first cruise ship to travel through the Northwest Passage was the *Lindblad Explorer*, an ice-strengthened Finnish-built vessel commissioned by pioneering adventure tour operator Lars-Eric Lindblad in 1969 as the first-ever ship custom built to take tourists to Antarctica. Ninety-eight passengers paid up to $20,000 for the privilege of being on the journey, which began in Newfoundland on August 20, 1984, and officially concluded when it reached Cape Lisburne, Alaska, on September 12. Aircraft from the Canadian Coast Guard monitored the vessel's progress, and an icebreaker broke through spots that the *Explorer* on its own could not; even so, the sense of accomplishment and adventure was clear. The ship's fifty (presumably all-male) crew resolved to begin growing beards at the journey's outset and did not reach for their razors until the last of the ice was behind them and the *Explorer* dropped anchor off Point Barrow—at which point, reported the ship's captain, Hasse Nilsson, "The passengers spontaneously broke out into celebration…so then we opened our champagne and toasted our achievement."

That achievement did not retain its uniqueness for long; the second cruise ship to transit the Passage, *World Discoverer*, did so the following year. In 1992, the *Kapitan Khlebnikov*—a powerful Russian icebreaker that routinely ferries passengers to both ends of the earth—made the first of, at time of writing, eighteen transits (the most of any ship). Twenty years later, the roster welcomed *The World*, a floating condominium containing 165 luxury residences continually circumnavigating the globe. It was all far removed from the comparative privations of the *Erebus* and *Terror,* the sailors aboard which would have surely been unable to even imagine a world that contained *The World.* Completely alien to them too would have been the immensity of the *Crystal Serenity,* the 2016 passage of which shone a spotlight on the region's changing conditions and generated fierce debate over the merits and defensibility of the previously impenetrable Arctic becoming a new frontier in tourism. Whereas the passenger ships that had previously transited the Passage generally offered somewhere between 100 and 200 berths or even fewer, *Crystal Serenity* boasted 1,070 passengers and 655 crew, making its entire complement larger than the communities it would be visiting along its journey. It was accompanied by an icebreaking escort vessel—the RRS *Ernest Shackleton*—carrying a helicopter and oil spill cleanup equipment; prior to the ship's departure from Seward, Alaska, representatives from Crystal Cruises met with the United States Coast Guard, the Canadian Coast Guard, Transport Canada, Alaska state emergency offices, and Alaska's North Slope Borough to confirm emergency response plans. The sheer size of the ship meant that the trip was criticized in some quarters, with then-director of Greenpeace UK, John Sauven, arguing that the "melting of the Arctic sea ice should be a profound warning for humankind, not an invitation to oil companies and now tour ships to move in." Despite the concerns, the voyage passed without great incident; the following year, *Crystal Serenity* repeated its journey, although it has not returned since.

Our own journey did not attract nearly so much attention; but then, with a maximum capacity of fewer than two hundred passengers and a tonnage just one-fifth that of *Crystal Serenity,* our profile was considerably smaller, literally and figuratively. Ours was not, however, an official transit of the Passage, as it had begun a short way inside the Passage's western entrance. The *Ocean Endeavour* had sailed from Greenland west into the Passage as far as the community of Kugluktuk, on the north coast of the Canadian mainland near the mouth of the Coppermine River; there, it disembarked its passengers, who would then fly on to Calgary, and welcomed us before journeying east. After a day of orientation and transit, we stepped ashore on an island randomly named after Jenny Lind, a nineteenth-century opera singer from Sweden; we marched across its low-lying terrain and squinted through binoculars and spotter scopes at a herd of musk oxen in the distance. Back onboard, we continued east, gathering on deck to watch a polar bear on a passing ice floe stained red with the blood of the freshly killed seal the bear was busily consuming. We reached the southern end of King William

Island and disembarked once more to visit Gjoa Haven, a community of perhaps 1,500 people, where we were greeted with a display of vigorous square dancing, introduced into the region by Scottish whalers who probed the icy waters of the Canadian Arctic Archipelago in the late nineteenth century but since adapted into its own unique form.

That Gjoa Haven even exists is a testament to the footprints left in the archipelago by Europeans—and in this case, one European in particular. He is the man who is memorialized by a cairn that stands on a hill overlooking the beach, who spent two winters here in the early twentieth century, who dubbed it the "finest little harbor in the world," and whose ship formed the basis of the community's name. It was during his sojourn here that he learned secrets of Arctic survival from the local Netsilik Inuit that stood him in good stead when he departed and that helped him become, a little more than a year after he departed, the first person finally to conquer the Northwest Passage.

Roald Amundsen is today arguably celebrated mostly for being the first person to reach the South Pole, in December 1911 (even if his success is, to some extent, overshadowed by the tragic circumstances of Robert Falcon Scott, the man whom he beat to the pole by a month and who died, along with his four companions, on the return voyage). But much of his life as an explorer was spent in the Arctic. He navigated the Northeast Passage and explored the ice of the Arctic Ocean on his ship *Maud* between 1918 and 1925; he was the first to fly over the North Pole, in an airship called *Norge*, in 1926 (and may have become the first person to achieve the North Pole by any means in the process, depending on the veracity of three previous and disputed claims to priority); and he met his death somewhere off the northern Norwegian coast while searching for another Arctic expedition. That the northern polar regions occupied so much of his biography seems entirely appropriate given that he was inspired to pursue a life of exploration when, at age fifteen, the young Norwegian came across Sir John Franklin's accounts of his overland expeditions in the Canadian Arctic and, as he recounted in his autobiography, *My Life as an Explorer*, "read them with a fervid fascination which has shaped the whole course of my life."

Of all the brave Britishers who for 400 years had given freely
of their treasure, courage, and enterprise to dauntless but

unsuccessful attempts to navigate the Northwest Passage, none was braver than Sir John Franklin. His description of the return from one of his expeditions thrilled me as nothing I had ever read before. . . .

Strangely enough the thing in Sir John's narrative that appealed to me most strongly was the sufferings he and his men endured. A strange ambition burned within me to endure those same sufferings. . . . Secretly—because I would never have dared to mention the idea to my mother, who I knew would be unsympathetic—I irretrievably decided to be an Arctic explorer.

After completing his military service and securing his Master's Certificate, he sought funding for, and establishment approval of, his goal. Fearing that exploration for exploration's sake would be insufficient, he dressed his plans in a scientific cloak, specifically a desire to reach the north magnetic pole and determine whether, as was theorized, it had moved since Sir James Clark Ross first located it in 1831. Having thus secured the initial backing he required, he procured a vessel, a forty-seven-ton sloop named *Gjøa* that had been plying its trade in the herring fishery off the Norwegian coast. It was, some advised him, far too small to win battles with the polar pack; but Amundsen had studied extensively what had and hadn't worked: "What has not been accomplished with large vessels and main force," he wrote, "I will attempt with a small vessel and patience."

After leaving port at midnight on June 16, 1903—under something of a shadow of secrecy to escape creditors—*Gjøa*, with its seven-person crew, made its way across the Atlantic and up the west coast of Greenland, turned west into Lancaster Sound, headed into the Passage, and on the evening of August 22 made its first stop, laying anchor just off Franklin's camp on Beechey Island. Amundsen recorded "a deep, solemn feeling that I was on holy ground, Franklin's last safe winter quarters." After one of their boats sank while transporting water to *Gjøa*, Amundsen noted that "the heaviness and sadness of death" hung over the island, and they moved on.

On September 9, with nights lengthening and the mercury dropping, they arrived at the south coast of King William Island in search of winter quarters. There, they found "the most beautiful little landlocked bay that the heart of a sailor could desire. High hills surrounding it on all sides would shelter us from the gales. Nothing could have been more ideal for our purpose, so the *Gjøa* was speedily brought inside, and we began our preparations for a permanent camp."

Within a matter of days, their camp attracted the attention of locals; before long, "fifty Eskimo huts sprang up about our camp, housing about two hundred men, women, and children."

Heavy ice forced the *Gjøa* to spend two consecutive winters in its temporary home, the Europeans learning Arctic survival from the Inuit, the two sides exchanging goods—and, it has long been rumored but never definitively established, the male crew members striking up relationships with some female Inuit. When he was finally able to set sail once more, Amundsen left behind what would become, with the arrival of a Hudson's Bay Company trading post in 1927, the permanent community of Gjoa Haven. Once more underway, *Gjøa* threaded through channels at times so shallow that there was barely an inch of water beneath the vessel's keel; for two weeks, Amundsen wrote, he "could not eat or sleep. Food stuck in my throat when I tried to swallow. Every nerve was strained to the limit in the resolve to foresee every danger and to avoid every pitfall. We *must* succeed!"

On the morning of August 26, 1905, Amundsen had come off watch and retreated to his bunk, only to hear running back and forth on the deck and then the triumphant words: "Vessel in sight!" Heading toward them from the west was the whaler *Charles Hansson* out of San Francisco. The quest was over. The young man who had been inspired by the writings of Sir John Franklin had accomplished what Franklin and so many others had died trying to do. He had completed the Northwest Passage.

"The glaciers are retreating. We used to have a nice one behind us, and that mountain over there used to have a nice cap on it, but we don't see nothing no more."

We had stepped ashore at Grise Fiord, on the southern tip of Ellesmere Island, where we were met by a young resident called Jessie Ningiuk, who was showing us around. As of the 2021 Canadian census, a mere 144 residents called this hamlet home; the only permanent public settlement on the tenth-largest island on Earth, it is one of the most northerly in the world and the farthest north in Canada. One of the coldest places to live on the globe, its Inuktitut name, Aujuittuq, translates as the "place that never thaws." Its record high temperature is only 60°F, and its daily mean across the course of a year just 10°F; the coldest temperature ever recorded was a bone-chilling

−52.6°F. Presented with such figures and asked for confirmation as to their accuracy, Ningiuk merely smiled and shrugged.

"Yeah," he said. "That sounds possible."

It is, however, not quite as cold as it used to be, and that has led, inevitably, to changes, and not just in the glaciers.

"We used to hunt caribou, but they just don't appear anymore," Ningiuk said. "I hope they just migrate on a different path now, and that they'll come back. But..."

"So, what meat do you eat now? Musk oxen?"

"Yeah." But then he paused and smiled. "Or mail-order caribou meat."

Mail arrives twice a week, delivered by a plane that lands on a makeshift airstrip in the shadow of the mountain that looks over the town. Once a year, in September, a barge brings the supplies—fuel, lumber, machinery, and other essentials—on which the community will function through the long, cold winter and beyond. A few times a year, small ships discharge groups of identically and brightly clad tourists to wander around, perhaps buy some souvenirs, and then return whence they came. Outside of such encounters, the people of Aujuittuq are on their own through extreme cold and mild warmth. If it seems an overly harsh and remote location for even Inuit to choose to reside, it was not initially much of a choice at all.

Larry Audlaluk, a community elder, explained as much to his visitors as they gathered indoors after their tour of the town. On this day, the community was in mourning; another elder, Peter Flaherty, had just passed away, causing the visit to be postponed for twenty-four hours. Thought had been given to canceling it entirely, but the community consensus was that it should go ahead because Peter always enjoyed greeting outsiders, and besides, theirs was an important story to tell. It was a story that began roughly 140 years before.

———— ∞ ————

On July 1, 1867, three British colonies—the Province of Canada, Nova Scotia, and New Brunswick—came together to create the Dominion of Canada, the former of the three devolving into two new provinces called Ontario and Quebec at the time of confederation. The new nation expanded rapidly: British Columbia joined in 1871, Prince Edward Island in 1873. A year before British Columbia's admission, Canada's geographic area expanded immensely as it annexed Rupert's Land (comprising modern-day Manitoba and much of Saskatchewan, as well as southern Nunavut and northern parts of Ontario and Quebec) and the North-Western Territory, both of which were de facto administered by the Hudson's Bay Company, which had been granted a charter for "sole trade and commerce" in the region by King Charles II in 1670.

But what is for our purposes the most germane stage in Canadian expansion was instigated almost by accident. On January 3, 1874, a Mr. A. W. Harvey, a British subject living in London, wrote a letter to the Under Secretary of State for the Colonies asking "whether the land known as Cumberland on the West of Davis Straits belongs to Great Britain and if it does—is it under the Government of the Dominion of Canada?" He would like to know, he continued, because he had been conducting a fishery there and anticipated erecting some temporary buildings. The following February, Lieutenant William Mintzer of the US Army Corps of Engineers wrote to the British consul in Philadelphia requesting a twenty-square-mile tract of land in Cumberland Gulf "for the purpose of carrying on a mining industry." This latter approach appears to have caused some kerfuffle in British colonial office circles, with one cable noting that "if this Yankee adventurer is informed by the British FO [Foreign Office] that the place indicated is not a portion of H.M. dominions he would no doubt think himself entitled to hoist the 'Stars and Stripes' which might produce no end of complications."

As correspondence on the matter bounced back and forth around the relevant governmental departments, it soon became evident that nobody could actually answer which Arctic islands and waters could be considered British territory and thus transferable to the new nation—that, in fact, "it appears that the boundaries of the Dominion towards the North, North East and North West are at present entirely undefined and that it is impossible to say what British territories on the North American Continent are not already annexed to Canada." Rather than assert the right to annex specific islands and waterways, and thus risk excluding those not yet explicitly claimed by British explorers or presently unknown to Europeans, it was proposed that London assert sovereignty over everything within a specific geographic area, the northern boundaries of which might be "the utmost limits of the lands towards the North Pole."

Several years of discussions and drafts confirmed the embarrassing truth that "the British did not know which of their arctic territories had not already been annexed to Canada, and that in any case an exact definition could not be given of territories that were then still largely unknown." As a consequence, the final proclamation that emerged was so broad as to be, without context, essentially meaningless, declaring that "From and after September

1, 1880, all British territories and possessions in North America, not already included within the Dominion of Canada, and all islands adjacent to any of such territories or possessions, shall (with the exception of the Colony of Newfoundland and its dependencies) become and be annexed to and form part of the said Dominion of Canada." Canada was, in other words, acquiring not just those parts of the Arctic archipelago that Britain's explorers had already claimed but also any other islands—claimed, unclaimed, and even unknown—that were vaguely in the vicinity. As the *Canadian Encyclopedia* subsequently observed, "This is a feeble basis for a claim of sovereignty."

Even so, Canada had, almost overnight, become the second-largest country on Earth; but initially it was an Arctic nation in name only, with no ability or perceived authority to enforce its vaguely drawn borders. British whaling fleets, which had been operating in Davis Strait since the 1820s, began expanding their operations into Lancaster Sound and the eastern part of the Northwest Passage shortly thereafter. A combination of steam-powered whaleships and a willingness to be frozen in over winter enabled them to probe more deeply into the Passage by century's end; but by then, a combination of their rapaciousness and the emergence of petroleum as a whale oil replacement for lighting and heating brought the industry to an end in the region. Similarly, by the time American whalers in the Bering Strait pushed north and into the waters of the western Canadian Arctic in the late 1880s, the clock was already ticking; whalers would ply those waters for only a few more years after the *Charles Hansson* encountered the *Gjøa* in 1905.

Of greater concern to the authorities in Canada was the ongoing presence of foreign explorers, particularly those who might be inclined to press territorial claims on behalf of their nations. Amundsen's voyage caused some consternation, but particular angst was prompted by a Norwegian named Otto Sverdrup.

In 1898, in command of *Fram*—a vessel on which Norwegian explorer Fridtjof Nansen had attempted to reach the North Pole and on which Amundsen would travel to Antarctica on his successful trek to the South Pole—Sverdrup set off for Greenland with ten crew members and five scientists to survey and round northern Greenland and explore and survey the unmapped northeastern part of the island. Prevented from penetrating any farther north as he probed the waters between Greenland and Canada, he parked the ship in the ice just off Ellesmere Island, and *Fram* would stay in that same area for a total of four years. The first Christmas was celebrated with "excellent food" and "rather too much to drink." The crew "danced to the melodies from a music box" and gave speeches "on several subjects, including both love and food." Within six months, however, tragedy struck when the doctor, Johan Svendsen, shot himself, leaving behind a letter explaining he could not bear life in the wilderness. Four months after that, another crew member, Ove Braskerud, also died and, like Svendsen, was lowered into the ocean through a hole in the ice. In May 1890, a spark from the galley started a fire on deck, destroying the mainsail, the ship's kayaks, several pairs of skis, and musk oxen and polar bear furs. The damage would almost certainly have been considerably worse, catastrophically so, in fact, had the crew onboard not been able to remove several boxes of gunpowder and had the fire reached a barrel containing fifty gallons of 96 percent alcohol.

Despite such calamities, however, and despite being unable to pursue the expedition's initial goals, Sverdrup turned the situation to his advantage, exploring and mapping the hitherto uncharted western part of Ellesmere Island; the northern coasts of Devon, Cornwall, and Graham Islands; and a trio of previously unmapped islands, which he named individually as Amund Ringnes Island, Ellef Ringnes Island, and Axel Heiberg Island and that are known collectively as the Sverdrup Islands—a total of approximately one hundred thousand square miles of land that, to the consternation of authorities in Canada, he claimed on behalf of Norway. Fortunately for those authorities, neither Norway nor Sweden, with which Sverdrup's homeland was at that point united, expressed a great deal of interest; nor was there a noticeable uptick of enthusiasm when Sverdrup offered up his claim anew upon Norway's independence in 1905.

Ottawa, however, was unaware of this lack of interest. Sverdrup's expedition, along with a series of voyages of varied degrees of success by Americans Adolphus Greely and Robert Peary, as well as Amundsen's completion of the Northwest Passage, prompted Canada to launch a series of periodic voyages into its Arctic Archipelago in order to reinforce its territorial claims: Baffin Island in 1897, Ellesmere Island in 1904, and, in 1909, a claim on behalf of the nation by Captain Joseph-Elzéar Bernier to the entire Arctic Archipelago, all the way to the North Pole.

There remained, however, uncertainty about whether land—perhaps even a continent—might lie beyond the archipelago, in the ice of the Beaufort Sea, a prospect that might severely complicate Bernier's claim—particularly if, for example, it bestrode the accepted boundary between Canada and Alaska.

That prompted the Canadian government to assume sponsorship of a planned expedition to the western part of the archipelago, abrogating the need for American backing, "expressly to forestall questions over any new lands that might be discovered." That expedition would be under the leadership of Vilhjalmur Stefansson, who had already prosecuted two voyages in the Canadian Arctic during which he had hired Inupiat guides (and fathered a son with an Inuvialuit seamstress), from whom he learned how to speak Inuktitut and survive in the region using only the supplies to be found there. What became the Canadian Arctic Expedition of 1913–18 soon experienced disaster when its ship, the *Karluk*, became trapped in the ice approximately halfway between Point Barrow on the northern coast of Alaska and Herschel Island, at the western opening of the Northwest Passage. A month later, in September 1913, with the ship still beset and with the prospect brewing of a long stay in the ice with dwindling supplies of meat, Stefansson set out across the ice with five others, including two Inuit hunters, to search for caribou and other game. Just four days after their departure, however, a blizzard pushed the *Karluk* and its ice floe to the west—away from the hunting party—at a rate of up to sixty miles per day; Stefansson and the *Karluk* would not see each other again. The ship continued to drift in the pack until it arrived about fifty miles north of Herald Island, off the northeast coast of Siberia; on the morning of January 10, 1914, it was crushed by the ice and sank.

Four men died while attempting to march across the ice to establish a camp on Wrangel Island, a larger landmass thirty-two miles to the west of Herald Island (their remains were found ten years later on the shores of Herald, which they had evidently reached by mistake). Another four died when, disillusioned with the leadership of *Karluk* captain Robert Bartlett, they attempted to strike out for Alaska. The rest were able to make camp on Wrangel, although two died of nephritis and one of a gunshot wound that may or may not have been self-inflicted, before a rescue party arrived on September 7, having been roused by Bartlett, who had marched across the ice to Siberia with an Inupiat hunter named Kataktovik before hitching a ride on a ship across the Bering Strait to Alaska.

When Stefansson returned from the Arctic in 1918 with news that he had found several new islands but determined that there was no substantial land north of the archipelago, he was feted for his discoveries by the National Geographic Society and the Explorers Club in New York but met an entirely more subdued reception in Canada. Although he would continue to write and lecture about the Arctic, and in 1921 conceived and promoted a highly controversial expedition to Wrangel, he did not return to the region again.

Perhaps surprisingly, between the efforts of Stefansson's group, the *Karluk* crew, and a southern party that operated somewhat independently of the rest, the Canadian Arctic Expedition achieved a number of its aims. In the words of the *Canadian Encyclopedia*, it "constituted a significant assertion of Canada's sovereignty . . . uncovered thousands of square kilometres of land and sea . . . made the last major territorial discoveries in the Archipelago . . . redrew substantial portions of the Arctic map, surveyed the coastline from Alaska to the Bathurst Inlet, and fixed errors in older charts of the region." It could not, however, entirely quell stirrings of external resistance to the notion of Canadian hegemony over the archipelago.

In 1920, in response to a Canadian request for Danish assistance in addressing reports that Inuit from Thule in Greenland (over which Denmark had authority) were hunting musk oxen on Ellesmere Island, Knud Rasmussen, a Danish ethnologist who had been living with the Inuit for eight years and was in charge of the trading post at Thule, replied that the section of Ellesmere where they were hunting was north of Canada and was a "no man's land." An official Danish cover letter that accompanied Rasmussen's response noted that the country's government "could subscribe to what Mr. Rasmussen said therein."

Canada filed a formal protest, and the government's anxiety only increased with Rasmussen's announcement that he was planning an expedition from Greenland across Canada to Alaska, which officials in Ottawa fretted would be used by Denmark to undermine Canadian sovereignty in the region. Ultimately mollified by Danish assurances that the effort would be purely scientific in nature, Canada nonetheless determined that it needed visible representations of its authority in the archipelago.

In the early years of the 1920s, Ottawa set up a series of RCMP posts: at Craig Harbour and Bache Peninsula on Ellesmere, at Pond Inlet and Pangnirtung on Baffin Island, and at Dundas Harbour on Devon Island, the site that those of us aboard the *Ocean Endeavour* would attempt to visit nearly a century later. In 1931, Norway officially renounced any interest in the Sverdrup Islands; in return, Canada paid Sverdrup $67,000 for his maps and journals. Alas, the old explorer died just fifteen days after the agreement was finalized.

———∞———

By the 1950s, an increased Canadian presence in the Arctic, and a growing awareness of and desire for the region's bounties, had transformed the lives of many Inuit in Canada completely from the nomadic subsistence that had characterized their lives prior to the arrival of outsiders. In the words of anthropologist Diamond Jenness, as quoted by writer Melanie McGrath:

> Hardly more than a generation earlier the Eskimos of the Central Arctic had produced their own food, their clothing, and all the other necessities of life, undisturbed by contact with the white man and totally ignorant of the infinite variety and complexity of his possessions. Then the storm had burst over them. . . . The pressure of the white man's demands, and his superior tools and weapons had shattered their self-contained existence; and his insistence on furs, ever more furs, had forced them to build up their lives on an entirely new foundation. Too weak to resist the aggressive and domineering invaders, they had surrendered their immemorial freedom and resigned themselves to the fate of helots laboring for foreigners who seldom troubled to learn their language. Now . . . their new economy too was tottering on its base, and it was by no means clear that the white man who had built the unstable edifice would not quickly abandon it again and leave the Eskimos to their fate.

Prior to the nineteenth century, interactions between white men and Inuit in the Canadian Arctic were sporadic, limited primarily to overwintering explorers and occasional interactions with the Hudson's Bay Company. That changed with the arrival of whalers in the mid-1800s, bringing Inuit and outsiders into regular, prolonged contact for the first time; and while those encounters were not without benefit for the Arctic's native inhabitants, the consequences were turbulent and profound. The whalers brought tools, from metal needles and harpoon heads to cooking pots, that they exchanged for furs; but they also carried diseases to which the Inuit had never been exposed, including tuberculosis, typhoid, measles, influenza, and diphtheria. As Melanie McGrath noted, "Entire families were wiped out by TB, whole settlements were ravaged by influenza. At the beginning of the twentieth century, the population of Southampton Island, around three thousand souls, was wiped out in a measles epidemic."

Some of the whalers were less interested in fur trading than in the attentions of Inuit women, and when they departed, they left behind syphilis and unclaimed children. And in their wake trailed missionaries, who brought rations with which they sought to alleviate the infanticide and elder suicide that were pragmatic but traumatic responses among Inuit to periods of scarcity, but who also foisted their own sexual mores on the Inuit, suppressed shamanism, discouraged drumming and other expressions of Inuit spiritualism, and sought to impose a belief system of their own.

Then the whalers left, but the twentieth century brought a new wave of arrivals, prompted by the likes of the Canadian Arctic Expedition and the RCMP, who, as part of Canada's assertion of presence and authority, sought to enforce in the Arctic the laws of those from far to the south. The role of the whalers was assumed by fur traders, and the Inuit progressively settled into communities adjacent to and increasingly reliant on the trading posts that provided them with cash and credit.

Decreased mobility did not correlate to increased stability. The Inuit were used to the ebbs and flows of the natural world, of periods of productivity being followed by fallower times; the vagaries of the economic system being layered on top of such uncertainty was not something for which they were well prepared. In 1943, a trapper on the Ungava Peninsula, that portion of northwestern Quebec that reaches out across the top of Hudson Bay, could command $35 Canadian for a good Arctic fox pelt; by 1950, with fashion trends changing in the lands to the south, that same pelt was worth one-tenth that amount. At the same time, the price of commodities in the north increased markedly, leaving many Inuit in an increasingly desperate situation and entirely dependent on welfare payments.

In 1952, a conference on Eskimo affairs convened representatives from government, the Roman Catholic and Anglican Churches, the Hudson's Bay Company, and even the National Film Board of Canada. Not invited: any actual Inuit, the reasons given being the logistic difficulties caused by transportation and language, as well as the fact that "it was felt that few, if any, of them have yet reached the stage where they could take a responsible part in such discussions." In their absence, the conference concluded that the solution to the "Eskimo problem" was to wean them off the cash economy and government welfare and that an ideal first step would be to relocate Inuit from places such as the Ungava Peninsula to the islands of the High Arctic Archipelago, specifically Ellesmere Island. Never mind that no Inuit had lived on Ellesmere for centuries, such are the extreme conditions of the High Arctic; the proposal's boosters expressed confidence that hunting would be plentiful and set about attempting to persuade the Inuit of that.

The promises and persuasion eventually bore fruit, and on July 25, 1953, the Arctic patrol ship *C.D. Howe* anchored off Port Harrison and prepared to pick up seven families, comprising thirty-one men, women, and children, plus their dogs and belongings. On August 28, the ship reached Mittimatalik, or Pond Inlet, on Baffin Island, where it picked up another three families, the idea being that these latter additions, accustomed to High Arctic conditions, would be able to guide the Quebec Inukjuamiut and help them settle on Ellesmere.

Little did any of them know that concerns had already been raised about the suitability of their destination. The squadron leader of the Royal Canadian Air Force base at Resolute, near where the Inuit would be discharged, expressed concern that "there was not sufficient wildlife in the Resolute area to sustain the Eskimo population," to which the reply was sent that, while there was reason to believe there was adequate marine life, "No one could say for sure that this was the case and, consequently, the experiment was being staged." Nor were the Inuit told until shortly before their arrival that, partly out of concerns over the impact of too large a settlement on local resources, they would be split up into three groups.

Upon arrival, it became immediately clear that the supplies that had been provided were wholly inadequate and inappropriate. At Resolute, they included twenty-four men's work pants in sizes 35 to 38, far too large for the Inuit men. There were two hundred gallons of gasoline "even though the Eskimos have no internal combustion engine of any kind." There were fingerless mittens. There were flimsy canvas tents but nothing with which to repair them. There was no first-aid kit, no oil lamps or rifles. Perhaps most egregiously, there was no sign of the three hundred caribou pelts that had been promised. There were a dozen buffalo hides, which they placed over the tents for warmth; but these made the tents dark, meaning that the women had to sit outside in the cold to do their work, and they were heavy; when snow arrived, they would have to be removed in case they might cause the tents to collapse.

The caribou skins that had been promised were the bare minimum of what would be required to equip all the families with clothing and sleeping bags. But there was no immediate prospect of making up the shortfall. The High Arctic Archipelago was within an Arctic game preserve that had been established in 1926. The Inuit were told they would be allowed only one caribou per family upon arrival, as hunting season had already closed, and although there was an abundance of musk oxen, that species was protected completely and shooting one was punishable by imprisonment or a fine.

There were other struggles: the temperature was far colder than they had commonly experienced, and the long darkness that descends on the High Arctic in winter, day after day, week after week, gnawed at their souls. Nor was there the abundant running water that they had taken almost for granted farther south, the families instead having to melt snow and ice. And the Inukjuamiut and the Pond Inlet Nunatsiarmiut eyed each other warily, the latter dismissive of the former's struggles to adapt and neither group able to communicate well with the other, so great were the differences in their dialects.

It appeared a genuine possibility that neither the experiment nor the families themselves would survive the first winter, with RCMP officers from Craig Harbour reporting that the Inuit were forced to burn wooden packing crates for warmth. The arrival of snow proved a turning point, as now they could at least fashion snow houses that would protect them from the elements; that winter passed, and another, and another. The settlers became more adept at hunting marine life and eking out a living in the unforgiving environment; once a year, the *C.D. Howe* returned, bringing much-needed supplies, and in 1955, it brought four more families from Inukjuak and two more from Mittimatalik. But even as the settlements developed, the transplants felt as if they were perpetually teetering on survival's edge.

After two years, several Inuit began asking to be returned home. Despite the promises they had been given when they were persuaded to leave their homes, they were refused. Samwillie Elijasialuk recalled that "they used to try to convince me not to go." Perhaps most pertinently, he was told, he said, that the "place has to be populated" and that if he wanted to return, he would have to find someone to replace him.

And therein, the Inuit subsequently contended, lay the truth about their relocation. While there unquestionably was a desire—born variously out of concern, compassion, and a wish to move impoverished Natives off the welfare roll and out of sight—to see the relocation experiment succeed for the reasons publicly offered, they argued that an underlying motivation was the belief that, as one official memo put it, "the occupation of these northern islands by Canada's first Arctic citizens only enhances our claim to sovereignty in these regions."

Concerns about sovereignty were, after all, the motivating forces behind the original decision to establish RCMP outposts at Craig Harbour and elsewhere in the 1920s and 1930s. They were also behind an earlier relocation attempt, when Ottawa reasoned that having established trading posts in the Arctic would be of greater consequence if there were people with whom to trade; and so, in 1934, fifty-four people from Baffin Island were moved to Dundas Harbour. The motivation for this move is explicit, a document in the government files stating that "to forestall any such future claims [to part of Canada's Arctic], the Dominion is occupying the Arctic islands to within 700 miles of the North Pole." (The Dundas Harbour trading post closed after a couple of years of poor ice conditions, after which some of the Inuit were returned home, while others were relocated across the Arctic as many as four times.)

Twenty years later, the greatest agita on this front was posed not by Norwegian explorers or Greenlandic caribou hunters (although the latter remained an irritant to Canada, and certainly a factor in the increased presence in the Arctic Archipelago, with Samwillie Elijasialuk reporting that the Inukjuamiut were "told to make the Greenland Inuit feel unwelcome"). Rather, the perceived danger to Canada's hold on its north came directly from its south: the United States.

During World War II, the United States established a military presence in Canada, occupying a string of airfields in the northwest for the defense of Alaska and building airfields at Churchill, Southampton Island, and Frobisher Bay in the east to ferry materiel to the European theater. While Canada had naturally assented to these moves, there was discontent over the fact that "Canadian authorities have little real say as to, for example, the exact placing of the airfields and the exact route of the roads on Canadian soil." Canada's Department of National Defence noted that it was "of great importance that Canada carefully safeguard her sovereignty in the Arctic at all times, lest the acceptance of an initial infringement of her sovereignty invalidate her entire claim. . . . Canada can no longer reasonably expect to maintain her Arctic territories in a state of vacuum, and hope at the same time to preserve her sovereignty over them *in absentia*."

These anxieties did not end with the cessation of hostilities. As global conflict segued into cold war, and the Arctic Archipelago became a de facto front line against a polar neighbor, Canadian officials noted with alarm continued US interest in the region. In 1946, Canada discovered that the United States was planning to build a series of weather stations in the High Arctic

Archipelago and that a US Air Coordinating Committee report recommended that US Army reconnaissance flights be conducted west of Greenland "to determine whether islands exist which might be claimed by the United States." Canadian authorities were able to persuade their counterparts in Washington, DC, to delay plans for the weather stations until 1947. The two countries subsequently established a number of such sites across the High Arctic together and, beginning in 1954, cooperated on a chain of more than sixty radar installations across the Arctic, from Alaska to Baffin Island, known as the Distant Early Warning Line, or DEW Line.

In anticipation of the latter, however, there remained nervousness about the fact that the great majority of the personnel involved in constructing and staffing such facilities would be American, as notes from a January 1953 Cabinet discussion make clear. The secretary of state for external affairs said that "it seemed clear that an increase in U.S. activity in the Arctic would present risks of misunderstandings, incidents, and infringements on the exercise of Canadian sovereignty" and that if "Canadian claims to the territory in the Arctic rested on discovery and continuous occupation, Canadian claims to some relatively unexplored areas might be questioned in the future." Prime Minister Louis St. Laurent added that it "was within the realm of the possible that in years to come U.S. developments might be just about the only form of human activity in the vast wastelands of the Canadian Arctic."

Later that year, the first wave of Inuit were relocated to Resolute and Grise Fiord—where, for more than thirty years, despite the promises that had been made to them, they stayed.

In June 1990, the Parliamentary Standing Committee on Aboriginal Affairs recommended that the government "acknowledge the role played by the Inuit relocated to the high Arctic in the protection of Canadian Sovereignty in the North" and that it apologize to the people of Grise Fiord and Resolute Bay and consider offering them compensation. In response, the government commissioned an independent report, which concluded that, in fact, not only was the relocation experiment "not motivated by a concern to strengthen Canadian sovereignty over the Arctic islands"; it was also "conscientiously planned," with the Inuit involved "benefiting from the experience."

In July 1994, the Royal Commission on Aboriginal Peoples held a series of hearings to investigate the issue, following which it concluded that "the precise extent to which sovereignty influenced the relocation is difficult to determine" but that there was "no doubt that the presence of Inuit settlements in the High Arctic, once established, did contribute to the maintenance of Canadian sovereignty." Whatever the motivation, the relocation "was not voluntary. It proceeded without free and informed consent. . . . Moreover, many Inuit were kept in the High Arctic for many years against their will when the government refused to respond to their requests to return."

Not until 2010 was an official apology forthcoming, the Canadian government expressing deep regret for "the mistakes and broken promises of this dark chapter of our history." By then, fifty-seven years had passed since the *C.D. Howe* made its way north with its first dispatch of passengers, and twenty-one years since a group of forty Inuit, mostly elders, were returned to Inukjuak—a move that had been three decades in the making and which in its delay caused a further rupture. For those who were adults during the relocations, the flights were a promise belatedly fulfilled and an opportunity to return home. But for those who were children then, who had grown up in the High Arctic and even been born there, Inukjuak was the place of mystery; it had taken thirty years, but now Grise Fiord and Resolute were home.

Among those who stayed was Larry Audlaluk.

"It was difficult," he said quietly after he finished telling his visitors from the *Ocean Endeavour* the story of the relocations and the birth and growth of Grise Fiord. "It was hard. But I grew up here. I had friends here. Family here."

He held our attention as he spoke, softly and calmly, with an occasional tear—telling us how his brother had been encouraged to come up as part of the second wave to be reunited with his father, only to find that his father had already died; how the first two years had been "simply survival"; how, for the youngsters who were part of the relocations, even amid all the suffering, sacrifice, and uncertainty, it had at times seemed like a "big adventure." He leavened the tale with moments of remarkable grace, noting that after Britain had signed over the Arctic Archipelago, the Canadians "did not know what to do with it. They did the best that they could."

As he concluded, he allowed himself the slightest hint of a beatific smile as he considered a community that had been born in deceit and despair but had survived to become, if not exactly thriving, then certainly strong.

"I don't hesitate to tell people what happened, but also to say, 'When you have difficult times, have hope,'" he explained. "There is always light at the end of the tunnel."

CHAPTER 3

Manhattan in the Northwest Passage

On December 26, 1967, a burst of gas shot out of a flow pipe on the north coast of Alaska, ignited, and erupted into a fifty-foot flame that burned for two days, lighting the Arctic night for miles. A little under three months later came the confirmation: Alaska was home to an oil field, and it was massive. Soon, that first well was producing almost 2,500 barrels a day, and a second, confirmation well was yielding even more: as much as 3,500 barrels daily. Covering almost a quarter million acres and holding an estimated 25 billion barrels of oil, it was the largest field ever discovered in the United States and remains one of the twenty largest found anywhere on the planet.

The discovery of the Prudhoe Bay oil field was not accidental. Inupiat Eskimos had noted oil seeps in the region at least as early as the 1800s and would use blocks of oil-soaked tundra for fuel. Late in that century, a US Navy exploratory expedition found oil in the area between what are now the towns of Utqiagvik and Deadhorse, and in 1922, geologist Alfred Brooks (after whom the state's northernmost mountain range is named) declared that the amount of oil that lay beneath the surface of what was then the Alaska territory would be "extensive."

But despite having had a century or more to prepare for the prospect of oil in Alaska, when the discovery came, those behind it were suddenly faced with the challenge of transporting it elsewhere. Alaska had been accepted into the Union only in 1959, and infrastructure was vanishingly thin on the ground. Certainly, there was no road that led from the relatively populated southern part of the state all the way to its northern coast. The state's governor at the time of the discovery, Walter Hickel, attempted to build one, granting permission for bulldozers to begin digging north through the tundra and permafrost, creating less of a road than an unusable water-filled ditch. The Hickel Highway, the project's supporters unofficially called it; more like the Hickel Canal, retorted its critics.

Of the three principal companies involved in exploring and drilling in Prudhoe Bay, two—British Petroleum and ARCO—favored the idea of transporting the oil via a pipeline running the length of Alaska south to Prince William Sound, whence it would be conveyed by tankers to their refineries on the West Coast. A third entrant, Humble Oil, had a different vision, motivated in part by the evident logistic difficulties of that plan but also by the fact that, unlike its two rivals, its refineries operated on the eastern seaboard. So Humble decided to experiment with another route, one that would bypass the Pacific coast entirely and bring the oil directly to its power center. It would try transporting oil through the Northwest Passage.

Roald Amundsen's transit of the Passage would be both the first in history and the last for forty years. On June 23, 1940, the *St. Roch*, a police schooner weighing a little over three hundred tons, measuring a shade more than one hundred feet from stem to stern, constructed almost entirely from Douglas fir and launched in 1928 as an Arctic patrol boat and support vessel for remote Royal Canadian Mounted Police stations, departed Vancouver and sailed toward Unimak Pass, the narrow passageway through the Aleutian Islands that acts as a gateway between the Gulf of Alaska and the Bering Sea. It paid a brief visit to the fishing port of Dutch Harbor on Unalaska; made its way into and north through the Bering Strait; rounded the northwest coast of Alaska; and pushed east into the Passage as far as ice would allow it before electing to spend the winter in a bay on western Victoria Island. It recommenced its journey in July 1941 but was ultimately forced by thick ice to spend a second winter farther east. It escaped this second icy harbor in August 1942, forced its way through the ice cover that continued to block its path, and ultimately arrived in Halifax, Nova Scotia, on October 11, becoming the second ship to traverse the Northwest Passage and the first to do so from west to east.

Officially, the purpose of the *St. Roch*'s voyage was to reinforce Canadian sovereignty in the Arctic Archipelago—and indeed, that was an important element of the journey's planning and of the *St. Roch*'s initial commission. The ship was not just to supply remote Arctic outposts for supply's sake but also to demonstrate in doing so that there was a permanence and plan to those outposts' establishment and operation. But expedition leader Henry Larsen also had other, secret, orders, not revealed for several decades.

When Nazi forces seized control of Denmark in June 1940, military strategists worried about what would happen to the Danish colony of Greenland—both in the sense of the strategic advantages the Axis powers might accrue from occupying an Arctic outpost, with deep coastal waters ideal for submarine harbors, so close to North America, and because the Allies' production of aluminum required a material called cryolite, the only source of which that was available to them being a solitary mine in a Greenland fjord. Britain and Canada drew up plans for landing a small occupation force to protect the mine, with the *St. Roch* to be deployed as a support and communications vessel.

In the event, Arctic ice put paid to both plans. Unable to force his way through the northern route offered by the Parry Channel, Larsen turned south, following in reverse the course pursued by Amundsen during his successful traverse. And during the second enforced winter in the Northwest Passage, Japanese planes bombed Pearl Harbor, and the United States, newly entered into the war, established its own military presence in Greenland.

Ultimately, reinforcing claims to sovereignty—which had been an initial goal of the expedition, then a cover for more secretive plans—became the overall rationale and reason for celebration of *St. Roch*'s journey, Larsen himself noting, "It was a great moment for me. Canada was at War and the Government realizing the need to demonstrate sovereignty over the Arctic Islands, was continuing to entrust the discharge of that responsibility to the Royal Canadian Mounted Police as it had done for decades."

Concern over the consequences for sovereignty of the temporary abandonment of the northernmost outposts persisted, however. The United States had established bases in the Canadian Arctic; what if the Americans chose to expand their reach into the highest reaches of the Arctic? Would they ask for Canada's permission, or, considering the area to be largely unoccupied and ungoverned, might they just move in? A further show of resolve and interest seemed opportune, and so the *St. Roch*, still in Halifax, was tasked with making a second attempt to navigate the Northwest Passage via a more northerly route, building cairns and leaving evidence of their progress along the way in order to lay down physical, as well as metaphorical, markers.

In the wake of the *St. Roch* came other firsts: the HMCS *Labrador* was the first warship to transit the Passage, as part of the first continuous circumnavigation of the continent; US Coast Guard ships *Spar*, *Storis*, and *Bramble* composed the first squadron, and were the first US-flagged ships, to make the journey; and the USS *Seadragon*, a nuclear submarine, traveled under the ice in 1960. But whereas Canadian anxieties persisted over sovereignty of the High Arctic islands of the archipelago, the Northwest Passage itself generally did not prompt much in the way of territorial defensiveness—until Humble Oil and the SS *Manhattan* set sail in August 1969.

The basic details of the ship and the voyage are these: Launched in 1962, the *Manhattan* was the largest merchant vessel ever to fly the United States flag and, at 115,000 tons, was significantly larger than most icebreakers being built at the time. Yet its fuel costs and crew requirements contrived to make it something of a white elephant, and it spent several years in a pedestrian existence, shuttling oil from the Middle East to the United States or carrying grain to India and Pakistan. When Humble wanted a vessel to attempt the Northwest Passage, the *Manhattan* seemed ideal, its twin propellers and 43,000 horsepower giving it the maneuverability and heft that the company felt would be required, not just of this voyage but of the future transits from Alaska to the Atlantic that it envisaged. In preparation for the journey, however, the *Manhattan* first had to be disassembled, cut into four sections, which were transported to different shipyards for upgrading to Arctic readiness: strengthening of the bow and of the hull, particularly around the engine room; installation of new propellers and protection for rudders; insulation of deck piping; addition of a heliport and upgraded communication and navigation equipment; and welding of a sixteen-foot-high, eight-foot-thick ice belt around the hull at the waterline.

The updated, strengthened, and improved *Manhattan* left Chester, Pennsylvania, on August 24, 1969, and with the help of some accompanying icebreakers, battled and bulled its way through the ice of the Northwest Passage to reach Prudhoe Bay on September 14. That done, it loaded a symbolic barrel of Prudhoe Bay crude oil and turned eastward back into the Passage, pausing to pay tribute to the graves on Beechey Island. It arrived in New York City on November 12—eleven weeks and two days after leaving Chester. But it had not emerged from its ordeal unscathed: in the final part of its journey, as it was heading east through Lancaster Sound, an ice floe gored a hole "big enough to drive a truck through" in one of the starboard tanks.

The image of the wounded tanker, coming just two years after the *Torrey Canyon* had spilled 120,000 tons of crude on the British coast after running aground, reinforced environmental concerns that had shadowed the ship since Humble announced its plans for the voyage. But before long, Humble had anyway concluded that the waters of the Beaufort Sea were too shallow to accommodate tankers of *Manhattan*'s size—let alone any larger, as Humble hoped to construct—and that building a terminal in that environment would not be feasible. The following year, Humble abandoned its plans for transporting oil via the Northwest Passage, and work soon began on a pipeline from Prudhoe Bay to Valdez.

(It was after leaving its eponymous port that the *Exxon Valdez* ran aground in 1989, emptying ten million gallons of oil into Prince William Sound. By then, the *Manhattan* was no more, its historic voyage giving way to a return to an unremarkable, itinerant life that ended when it was severely damaged in a typhoon while anchored off Yeosu, South Korea, in 1987; with repairs considered too expensive, it was shuffled off to a scrapyard in China, where it met its unrecorded end.)

But if the outline of the journey itself was relatively straightforward, the emotion surrounding and reaction to it were anything but. The announcement of the *Manhattan*'s voyage, and the revelation that the tanker would be accompanied by the US Coast Guard icebreaker *Northwind*, prompted fulminations in Canada's media and Parliament, their particular angst focused on the fact that not only had the United States not requested permission to enter the Northwest Passage, but it had also pointedly refused to do so. From the American perspective, this was nothing especially notable or provocative: it was consistent with existing policy and prior practice. While Washington might have recognized Canada's dominion over its Arctic territories, it regarded the Northwest Passage as an international strait through which peacetime travel could not be encumbered; to request permission to travel would be, as the United States saw it, contrary to its rights and requirements under accepted international maritime law. And the *Manhattan* was hardly the first US-flagged vessel to make the trip; the *Spar, Storis*, and *Bramble* had done so in 1957, after all, without generating too much fuss.

To its critics, however, the *Manhattan*'s voyage was an altogether different animal. This was a commercial vessel, not a military one, and it potentially presaged the arrival of many more, all of which—were the US position to be universally accepted and adopted—could conceivably be making the trip without legal restraint. Not only that, but many or all of them could be carrying oil through Canada's Arctic, a prospect that fed directly into the anxieties of the nascent environmental movement, particularly in light of the *Torrey Canyon* disaster—anxieties that would be further stoked by the breakup of a Liberian tanker off the Nova Scotia coast in 1970.

The initial response of the government of Prime Minister Pierre Trudeau to the prospect of American tankers steaming through the Northwest Passage and undermining Canadian sovereignty was, perhaps surprisingly, somewhat sanguine. On May 15, 1969, Trudeau told the House of Commons that the "legal status of the waters of Canada's Arctic archipelago is not at issue" as a result of the *Manhattan* voyage. "The Canadian government has welcomed the *Manhattan* exercise, has concurred in it and will participate in it," he continued.

Defenders of the government's relaxed position underlined that both the US Coast Guard and Humble Oil had actively sought input from, and the cooperation of, Canadian authorities; that the United States had specifically requested Canadian coast guard vessel support and accompaniment; that the *Manhattan* intended to fly the Canadian flag when appropriate; and that in return for Canadian cooperation, Ottawa would receive all the ship and ice data that accrued during the voyage.

But throughout the postwar period and deep into the Cold War, Canadian policy on Arctic sovereignty had been predicated on preventing American expansion into the archipelago, and having been conditioned to see the issue through that prism, the country's media were not about to let go of that bone now—and all the more so at a time when, with war in Vietnam raging and the United States roiling as a result of racial and social upheaval, Canadians were regarding their southern neighbor with rather more hesitant and anxious eyes. And the fact that the United States did not and would not ask for permission to enter the Passage was a stick with which journalists continued to beat the government, no matter what assurances Ottawa issued that the situation was under control. The *Manhattan's* voyages "made Canadians feel they were on the edge of another American . . . [theft] of Canadian rights and resources"; if other tankers followed the *Manhattan* through the Passage's waters, then Canada "must be ready to receive and control them." Sensing an opportunity to make mischief, the Soviet Union weighed in, with *Pravda* editorializing that "the U.S. military has been rapidly encroaching on the sovereignty" of Canada through its Arctic activities. One member of Parliament claimed that US maps "indicated U.S. ownership of potentially oil-rich islands in the Archipelago." Others chartered a plane and flew over the *Manhattan*, calling it over the radio to welcome it into Canadian waters, while yet others joined the HMCS *Labrador*, the Canadian icebreaker accompanying the tanker, and from there boarded and were given a tour of the Humble Oil vessel.

Given the growing furor, the Trudeau government felt increasingly compelled to find some way to assert the country's authority, even as it sought to avoid stoking an unnecessary conflict with Washington. The solution it came up with was to focus on the very real concerns about the possible impact of oil exploration and transport on Canada's Arctic waters and shorelines. In August 1969, the government announced that it would introduce legislation setting out measures necessary to prevent pollution in the Arctic seas, and the following April, it made good on that promise.

One bill, the Arctic Waters Pollution Prevention Act (AWPPA), created a zone of 100 nautical miles (115 statute miles) within which Canada had authority to enforce pollution prevention measures, while another amended existing legislation to extend the limits of the country's territorial waters from three miles from each coastline to twelve, largely in line with the path being followed by other coastal nations.

One function of the second piece of legislation was to create a de facto backdoor assertion of authority over the Passage, in that there were choke points in the Passage that were less than twenty-four miles wide, indirectly supporting the Canadian claim that the Northwest Passage was not an international strait, thus subject to Canadian regulations. Some of the regulations that were ultimately spelled out under the AWPPA included a requirement that ships entering the Passage conform with Canadian standards; vessels failing to meet those standards could be halted and their cargo seized.

Ottawa was keen to thread the needle: to assuage domestic critics and produce meaningful legislation to address legitimate concerns while also not outwardly provoking conflict with the United States and, importantly, staying within an international consensus. "It is not an assertion of sovereignty," Trudeau emphasized of the AWPPA; "it is an exercise of our desire to keep the Arctic free of pollution."

Nonetheless, the United States reacted with hostility. When advised of the impending legislation, President Richard Nixon announced that the United States would reduce its import of Canadian oil. Congress authorized construction of "the most powerful ice-breaker fleet in the world." If the United States did not oppose Canadian action, "it would be taken as precedent in other parts of the world for other unilateral infringements of the freedom of the seas."

Such opposition only hardened Canadian resolve. "The United States objects to the Canadian bill," thundered the *Globe and Mail*. "Let it. . . . This issue has a significant potential for confrontation between our two nations. But Canadians should realize we are different nations with different interests and different purposes. . . . And so Canadians should begin to prepare for a moving apart, not only on this issue, but many others affecting our economy, our culture, our approach to international affairs."

Even with the idea of using the Northwest Passage to transport Alaskan oil abandoned in favor of a pipeline, Canada proceeded with development of the AWPPA, formalizing it with detailed regulations on the icebreaking capabilities, navigational procedures, and even construction of vessels that entered the Passage, and following that up in 1977 with a voluntary reporting system called NORDREG, under which vessels over three hundred tons were requested to submit a clearance request, including route and destination, twenty-four hours prior to entering the Passage and to file daily updates with the Canadian Coast Guard Traffic Center during their journey.

Meanwhile, Canada continued to seek international authority for its position, pursuing as a primary avenue the United Nations Convention on the Law of the Sea (UNCLOS), the third negotiation session of which, dubbed UNCLOS III, began in Venezuela in 1973. There were three issues of particular relevance to Ottawa: the status of the Arctic Ocean and of the Northwest Passage and the ability of Arctic states to regulate pollution in their coastal waters. The results were, however, a mixed bag from the perspective of Canadian concerns: by the time UNCLOS entered into force in 1982, it did contain a provision that Arctic states indeed had the authority to establish and enforce measures to control coastal pollution; however, in order to achieve consensus on that issue, it had to grant concessions on related ones, so that when all was said and done, UNCLOS provided Canada only with the authority to impose limitations on nongovernmental ships. Worse from a Canadian perspective, the right of "innocent passage" through international straits would be replaced by one of "transit passage," which effectively granted ships the same right as on the high seas—the suite of compromises meaning, noted Shelagh D. Grant in her book *Polar Imperative*, that Canada could exercise control in the Northwest Passage only "over non-governmental ships that violated international anti-pollution standards."

But then, in 1985, Canada and the United States were confronted with a *Manhattan* situation redux in the form of the US icebreaker *Polar Sea*. The *Polar Sea* was based in Seattle but had temporarily assumed responsibility for resupplying the Americans' Thule Air Base in Greenland; the US Coast Guard suggested that it should steam through the Northwest Passage rather than detour all the way through the Panama Canal. Canadian officials were apprised of the plan, and all sides agreed that the *Polar Sea* should carry three Canadian observers and that the CCGS *John A. Macdonald*, which had accompanied the *Manhattan* on its return voyage, would do likewise with *Polar Sea* in areas of heavy ice.

Both sides agreed that the journey and its attendant arrangements would have no impact on their divergent stances regarding the status of the Northwest Passage; but if all involved assumed that this meant the voyage would unfold without event, they either had forgotten or were somehow unaware of the *Manhattan* controversy. Sure enough, once news of the impending expedition reached Canadian media and opposition politicians, and specifically the fact that explicit approval for it would be neither procured nor proffered, a collective national angst erupted anew.

In response, the administration of Prime Minister Brian Mulroney committed to building an icebreaker four times more powerful than anything presently available for it to deploy, to acquire up to a dozen submarines, and to develop a sonar system to monitor foreign submarine activity (none of which actually transpired). It also was finally moved to do what it had previously been somewhat resistant to do: it drew a boundary line around all the islands in the Arctic Archipelago and declared that all the passages contained therein were internal waters.

That move prompted the United States neither to acquiescence nor to outrage; but three years later, the two countries did sign a bilateral Arctic cooperation agreement, under which they would effectively agree to disagree over the Northwest Passage status, but which included a commitment by the Americans that "navigation by United States icebreakers within waters claimed by Canada to be internal would be undertaken with the consent of Canada." The first test of the arrangement took place in October 1988, just nine months after the agreement was signed, when American authorities, while underlining that they were not acknowledging any Canadian claim to sovereignty, requested permission for the USCGC *Polar Star* to traverse the Passage. The Americans assured the Canadians that the vessel would adhere to the AWPPA and other applicable Canadian laws and that any necessary pollution control, mitigation, or cleanup would be the responsibility of the United States. Ottawa assented, and the voyage passed without incident.

With that, a truce effectively descended on the Northwest Passage and its sovereignty, one that more or less persists. But thirty years later, it would experience its first real test in decades as Washington and Ottawa began sniping at each other over the issue anew. The tiff was largely due to the confluence of an especially bilious administration in the United States and an atypically assertive and nationalistic one in Canada, but the seeds had been sown in a most unlikely setting: the graves on Beechey Island.

CHAPTER 4

The Faces of Franklin

For 138 years, John Torrington had lain undisturbed. Now, on August 17, 1984, he was about to reemerge, the earth and permafrost above his coffin dug away to reveal a stout mahogany coffin, adorned with a lovingly hand-painted plaque detailing the first member of the Franklin expedition to meet his demise: "John Torrington died January 1st, 1846, aged 20 years." The wood on the coffin lid was too soft to pry open without it breaking apart; for the same reason, those assembled elected not to pull the nails that had fastened the lid to the box. Instead, using a chisel, they carefully sheared through the nails with a chisel head and gently lifted the lid. Inside the box, Torrington's body lay entombed in a block of ice that had built up over the decades; using both hot and cold water, they gently melted the ice away and found themselves face-to-face with his remarkably well-preserved body. Torrington's eyes were open, as if staring blankly while in mid-thought, his lips pulled back, exposing his teeth. Torrington was slight—just five feet four inches tall—and had been buried in gray linen trousers and a white, blue-striped shirt.

Two of those present lifted him gently out of his coffin: one, Arne Carlson, lifted his legs and another, expedition leader Owen Beattie, supported his shoulders and head:

> As they moved him his head rolled onto Beattie's left shoulder; Beattie looked directly into Torrington's half-opened eyes, only a few centimeters from his own. He was not stiff like a dead man, and although his arms and legs were bound, he was limp. "It's as if he's just unconscious," Beattie said.

They lowered his body gently to the ground, slowly undressed him, and carefully began an autopsy.

In the century and a half since the *Erebus* and *Terror* disappeared, more evidence had been gathered that shed some light on the events of 1845 and beyond, while still leaving much of the story in the shadows. Explorers including Charles Francis Hall and Frederick Schwatka had returned to King William Island in the decades after Francis McClintock and William Hobson discovered bodies, artifacts, and the cairn with the note detailing the death of Sir John Franklin. They had gathered much more Inuit testimony, including claims of grave sites, and over the years, many others had sought to follow up on and expand their findings. Among them was Beattie, a forensic anthropologist at the University of Alberta, who in 1981 led a small team to the island to search for a grave, the approximate location of which Inuit had pointed out to Hall but which he had been unable to pinpoint because of heavy snow.

There, they found scattered bones and part of a skull from a Caucasian male, many of them showing pitting and scaling on their surface, indicative of scurvy. Examination of what Schwatka had believed to be a Franklin expedition grave site found bones of what turned out to be an Inuk male; two more sites also proved to be of Inuit origin. Beattie and his team took samples for analysis and returned home.

Initial examination of the Caucasian skeleton provided apparent confirmation of the most contentious aspect of Inuit testimony—later bolstered by additional research on other skeletal remains from the expedition, including from a previously unknown site discovered in 1992 that contained the bones of at least eleven men. Parallel grooves in a femur were indicative of knife marks; subsequent analysis by Beattie and Simon Mays, a human skeletal biologist for Historic England, found evidence of breakage and "pot polishing": a smoothing that can occur when the ends of bones rub against a cooking pot when they are placed in boiling water. The verdict was unequivocal: not only had the survivors eaten the meat of their fallen comrades, but they had also resorted, in desperation, to boiling their bones to extract the marrow.

Another discovery, however, came out of the blue. The Caucasian bones, in contrast to the Inuit remains, contained extremely high levels of lead— at 228 parts per million, well above the minimum levels required to cause lead poisoning. Was it possible that this had contributed to, or even directly caused, the failure of the Franklin expedition? And if so, whence might such high levels of lead have come? The primary suspect was the eight thousand cans of food that were loaded onboard *Erebus* and *Terror*, which were sealed with lead solder. Certainly, in the aftermath of the ships' loss, some of those casting around for blame had focused on the process, newly developed, by which the food was canned, although in this instance the accusation was that the food had been improperly prepared and contained botulinum toxin. Perhaps, Beattie ventured, the expedition members realized the food was killing them, and this was behind the otherwise seemingly inexplicable decision to establish piles of empty cans on Beechey Island. To bolster his case, he sought and was granted permission to exhume the graves on Beechey, study of which confirmed that those corpses, too, contained highly elevated levels of lead.

Subsequent research has cast doubt on Beattie's claims: first by suggesting that the amount of lead in the solder was insufficient to account for the levels detected in the bones and tissue, and proposing that a likelier culprit was the lead pipes in the ships' new fresh water–making system, and then by concluding that the levels were merely indicative of the extremely high amounts of lead to which many people in Victorian England were exposed. Perhaps high amounts of lead in the body, when combined with all manner of deprivations suffered by the crew, contributed to their weakness and demise, but it was not, it is now believed, a primary factor.

The most significant consequence of Beattie's work, however, lay not in his tissue analysis but in the photographs he and his team took of John Torrington—and of John Hartnell and William Braine, whom they also exhumed. The enduring aspect of the Franklin expedition had been the mystery surrounding it, the paucity of tangible evidence of the crews' fate, the absence of the ships. Now, suddenly, people around the world could and did stare with wonderment at the visages of the trio who had died on Beechey Island and look upon the mummified expressions of young men who almost looked as if they could be reanimated. The expedition, which had begun to fade from widespread memory, was suddenly front and center once again. It inspired a new wave of literature and music, from the predictable (American folk singer James Taylor's "The Frozen Man") to the implausible (British heavy metal group Iron Maiden's "Stranger in a Strange Land"). Notably, however, for the first time the expedition became a Canadian story.

Canadian interest in Franklin, and consideration of him as a figure in Canada's history as well as Britain's, was arguably first prompted by the 1981 release of the song "Northwest Passage" by Stan Rogers, one of the country's most popular singers. The song's lyrics emphasized the geography of

Franklin's search, referencing iconic Canadian place-names such as Davis Strait and the Beaufort Sea and drawing a parallel between the expedition and a modern-day journey across the nation's prairies. In 1995, in a contest on CBC Radio to pick an "alternate national anthem," "Northwest Passage" was an overwhelming choice.

But Beattie's photographs, and their inclusion in a 1987 book that Beattie wrote with John Geiger called *Frozen in Time*, elevated the expedition into the national consciousness to an unprecedented degree. Margaret Atwood, perhaps Canada's most celebrated and cherished contemporary author, contributed the book's foreword and featured Franklin in her own writings. In 1992, Canada designated the undiscovered remains of the expedition ships a national historic site; five years later, Britain agreed in principle to transfer ownership of the wrecks to its former colony.

The political situation surrounding Canada's Arctic and the Northwest Passage had returned to a quiescent state that stood in contrast to the fevered controversies surrounding the *Manhattan* and the *Polar Star*. That state of affairs surely pleased officialdom in Ottawa and Washington but agitated observers such as author John Honderich, who asked rhetorically in 1987 whether Canada was "losing the North" as a result of its "benign neglect" toward the Arctic.

Those seeking a more muscular approach to Canadian Arctic affairs, however, finally got what they wanted in 2006, with the election of Stephen Harper as prime minister. Shortly after taking office, Harper asserted that "Canada must do more to defend Canada's Arctic sovereignty." In doing so, he said, "we are not only fulfilling our duty to the people who called this northern frontier home, and to the generations that will follow; we are also being faithful to all who came before us." He developed a four-pronged Northern Strategy that placed "exercising sovereignty" at the top of the agenda. He pumped funding into Canada's northern regions, including for the country's military. Perhaps more than that, argued Rob Huebert of the University of Calgary, he sought to "integrate the Canadian north into the nation's psyche."

At the heart of this latter goal was his attachment to Canada's Arctic history and in particular the Franklin expedition. Described as being generally reserved in public settings, Harper could be at ease and even animated when discussing Franklin and stories of other Arctic expeditions. John Geiger described it as a "deep personal interest." Others offered that he was "obsessed."

All of which helps explain why, in 2008, as the world was entering its biggest financial crisis in eighty years, his government announced a bold new venture.

Canada would find the *Erebus* and *Terror*.

CHAPTER 5

Return to the *Erebus*

It was morning. The ship was not moving.

I blinked awake and stared briefly at the ceiling. The soft glow of the television monitor in the corner warmed the cabin; the map it displayed revealed that we had reached the day's destination. I cantilevered my torso half out of my bunk and peered through the window for visual confirmation.

The scene, out of context, was unremarkable. The Arctic water was slightly choppy, the sky gray, the two melded almost together but for a barely noticeable interruption from a nondescript strip of low-lying land. In the near distance sat a ship at anchor and, near to it, a barge.

I stared at the barge with an intensity that, to all outward appearances, it did not deserve. I stared not so much to see the barge more clearly as to open a portal to the past, to project onto our surroundings the scene approximately 170 years before, to imagine the brutal conditions that had brought an end to a miserable four-year expedition and ultimately led to all of us—the barge, its companion vessel, the ship I was on, its other passengers, me—to this place. I stared not at the barge itself but at a mental image of the ship that had sat in its very spot and that now lay directly beneath it, resting on the seabed in eleven meters (thirty-six feet) of water.

Within a few hours, I would be in that same spot. I climbed into one of the Zodiacs that was ferrying groups back and forth, and we headed away from the *Ocean Endeavour* and toward the ship, the barge, and the wreck of HMS *Erebus*.

As we clambered onto the barge, we were met with a wall of sound from loudly clattering pumps delivering air to the divers who at that moment were exploring the wreck below. One of the team, Brandy Lockhart, a Parks Canada archaeologist who herself had dived on the wreck, took a few of us to a small room to peek at some of the artifacts that she and her colleagues had retrieved.

We were the first to see them—we were in fact the very first people, outside of the archaeological team, to visit the wreck site; afterward, she explained, the artifacts would be taken to Gjoa Haven for display in the community and then sent to Ottawa for analysis and preservatory treatment. Having been immersed in salt water for 170 years, they needed to be kept wet, as they would not survive rapid drying. Accordingly, they were in shallow tubs of water and covered with damp towels.

Brandy gently uncovered the items in the first tub: a pair of small bottles, one ceramic and one glass, which at some point may have contained liquor—and, perhaps, did still. In the next tub, the leather sole of a shoe. In the next, what appeared to be sugar tongs.

They were striking in their ordinariness; and yet, their very mundane nature made them feel all the more remarkable. One of the officers had presumably used the tongs to casually drop a lump of sugar into some tea, perhaps while wearing the very same boot that now lay in tatters before us. Perhaps at the time, he had been discussing the expedition ahead, or his plans upon his return to England; subsequently, he would have grown ever more pessimistic and despairing, ever weaker and more desperate, before ultimately dying in misery, most likely somewhere on King William Island nearby.

We left the room and returned to the deck of the barge, where a pair of operators looked at a monitor showing a live feed from the diving operation presently underway. Mike Bernier, the project manager, explained that the wreck lay partly beneath and off to the side of the barge; from where I was standing, the bow of the *Erebus* was over my right shoulder. The diver whose feed we were watching was, explained Mike, exploring officers' quarters on the port side of the ship.

"Right now," he said, "he's looking at some china," and as he spoke, the diver trained his camera on a stack of blue-and-white plates that looked for all the world as if they had been neatly stacked just the day before.

I walked to the railing and peered over. Directly beneath me lay what remained of HMS *Erebus*. How did it get here, so far from the location it had, according to the note the crew had left behind, been abandoned? How did it sink? When? What were conditions like when it disappeared beneath the ice and waves? Was anyone still onboard? Were any of the expedition members still alive? What secrets did it clasp within its slowly disintegrating hull?

We had hoped to look at the wreck directly by lowering a bathyscope into the water. As soon as I was advised of that possibility, I had become fixated on it, obsessed with the prospect of viewing the *Erebus* with my own eyes.

Alas, the turbidity in the water precluded us from doing so, and while I was initially disappointed, I soon put that to one side. I was almost within touching distance of the famous HMS *Erebus*. More than that, I was standing where it had died. It was enormously exciting but also profoundly sad.

In August 2014, Stephen Harper stood on the bridge of the HMCS *Kingston* as it steamed from Pond Inlet to Arctic Bay in the eastern reaches of the Northwest Passage. The *Kingston* was participating in the search for the *Erebus* and *Terror,* a Harper-mandated quest that was now in its sixth year; despite the lack of success to this point, however, Harper was bullish about its prospects.

"You've heard what I've said before about that, eh?" he mused to his accompanying press pool. "That one day we're going to come around the bend, and there's going to be the ship, and Franklin's skeleton slumped over the helm. We're going to find it."

His professed optimism was not at that moment shared by those conducting the search. That summer was proving to be the heaviest ice year since Harper set the renewed hunt in motion, and the search team was uncertain whether conditions would change sufficiently to enable them to make much headway. And yet, within just a matter of weeks, Harper was making the announcement he had been dreaming of. On September 7, 2014, a remotely operated underwater vehicle, launched from the icebreaker *Sir Wilfrid Laurier,* had found the HMS *Erebus*; two years and five days later, searchers would announce the discovery of the *Terror*.

"This is truly a historic moment for Canada," Harper declared. "Franklin's ships are an important part of Canadian history given that his expeditions, which took place nearly 200 years ago, laid the foundations of Canada's Arctic sovereignty." Never mind that the *Erebus* and *Terror* had been British ships with British crew, seeking to expand the influence and reach of the British Empire, or that London's transfer of North American Arctic sovereignty had been a somewhat muddling affair that was set in motion several decades after Franklin's disappearance; the entire endeavor, by Harper's telling, had been an iconic Canadian adventure.

"The discovery of two historical wrecks from the 1840s that sailed under the authority of Britain before Canada was even a state doesn't really extend our claims of control over the waters of the Northwest Passage," noted Rob Huebert of the University of Calgary. But, he added, "the search for the

Franklin [ships] allows us to have a metaphor as we develop technologies that do, in fact, allow us to asset better control over the region."

Added Carleton University's Jeff Ruhl: "Harper's use of Franklin has served as a way in which all the different elements of his northern agenda can be packaged in a singular thing. Franklin has served the project in many ways as the symbolic manifestation of Harper's overall northern agenda."

All of which sounded decidedly vague and open to interpretation. But what exactly was Harper's goal for the Arctic—and specifically the Northwest Passage? What was the Arctic policy he espoused? How did that fit with Canadian policy that both preceded and superseded his administration?

———∞———

For all their initial muscularity and focus on sovereignty and might, Harper's Arctic policies became more nuanced and moderated over the years in response to changing circumstances regionally and internationally—and, as noted in a 2016 analysis for the Canadian Forces College, "as the government learned more about the North through its time in office."

The aforementioned Northern Strategy, for example, while heavy on issues of sovereignty, was broader and softer in tone than the "Canada First Defence Strategy," a document released the year before, in 2008, which emphasized that Canada's armed forces need to "have the capacity to exercise control over and defend Canada's sovereignty in the Arctic." The Northern Strategy also addressed issues such as "promoting Social and Economic Development, protecting our Environmental Heritage, [and] improving and Devolving Northern Governance," and even on the issue of sovereignty, it offered reassurance that any disagreements with the country's fellow Arctic nations were "well-managed and pose no sovereignty or defense challenges for Canada." It also quietly dropped the mantra of Harper's first few years as prime minister, that "the first principle of Arctic sovereignty is: Use it or lose it," dialing back on the notion that unless Canada showed a forceful and expansive commitment to the Arctic, its territorial claims might come under threat. Another document, "Statement on Canada's Arctic Foreign Policy," issued in 2010, continued this subtle shift from a purely utilitarian and unilateral approach to the region, focusing more on Arctic governance and underlining the need to find "ways to work with others: through bilateral relations with our neighbors in the Arctic [and] through regional mechanisms like the Arctic Council."

This trend has continued following the replacement of the Harper administration with the government of Justin Trudeau, whose 2019 Arctic policy identified eight priorities with very different emphases from those of the early Harper years. While it states that the Canadian Arctic and its peoples should be "safe, secure and well-defended," it also makes explicit reference to the "rules-based international order in the Arctic" and prioritizes the health of both people and the environment in Canada's northern regions.

Some of Harper's specific policy promises regarding securing Canada's Arctic sovereignty also fell by the wayside or were downgraded or delayed. For example, a commitment to build three armed heavy icebreakers "capable of operating anytime in the North at any time of year" morphed into the construction of smaller Arctic offshore patrol ships, which were derided by some as "inadequate slush-breakers" but in fact are both more versatile and more appropriate to a broad-based approach to the region in that they can also act as fisheries patrol and swift disaster response vessels. (In July 2024, Ottawa announced it would commission a dozen conventionally powered submarines capable of operating under Arctic ice "to ensure that Canada can detect, track, deter and, if necessary, defeat adversaries in all three of Canada's oceans.")

There was a mixed follow-through on a 2007 Harper promise to "install two new military facilities in the Arctic to boost Canada's sovereign claim over the Northwest Passage and signal its long-term commitment to the North." One, the Canadian Armed Forces Arctic Training Centre in Resolute Bay, Nunavut, is up and running, offering a "permanent location for training and operations in the High Arctic." A proposed naval refueling facility in the northwest of Baffin Island, however, did not break ground until 2015, was expected to open in 2018 but did not, has seen elements such as an airstrip removed as the budget has ballooned, saw construction halt during the COVID-19 pandemic and a road connecting the base to the community of Arctic Bay washed out, and is, at time of writing in summer 2024, now projected to be operational by the time this book is published.

If the realities of Canadian Arctic policy under Harper didn't quite reflect the initial tub-thumping, then a growing understanding of the complexities and nuances of the issue of Arctic sovereignty may well be, as the Canadian Forces College analysis expressed, a primary reason. It is also likely a reflection of the fact that actual policy, on any issue or in any country, regularly ends up being a diluted version of the red meat offered to supporters as budgetary and bipartisan realities bite.

But the timing of the retrenchment also suggests that it may owe something to the fact that the initial posturing may have proven more effectively provocative than anticipated. In 2007, a year after Harper was elected, a Russian research ship deployed a submersible that dropped a Russian flag on the seabed at the North Pole, an act that might to some extent have been prompted by Canada's saber-rattling and which, as we shall see later, almost immediately led Arctic states to dial down the rhetoric and pledge to work together in the region. One result of that Russian action was to drive Canada, which had generally been supported on the issue of the Northwest Passage by Moscow, to align itself more with the Arctic policies of the United States, opposition to which had been the cornerstone of Canadian sovereignty concerns since Britain yielded control of the North American Arctic more than a century and a quarter previously.

As Joël Plouffe observed in a 2014 analysis for the Canadian Global Affairs Institute, in "his first year in power, the Prime Minister often suggested that the most obvious threat to Canada's sovereignty was its nearest neighbor and closest ally, the United States," but the 2010 statement on Canada's Arctic foreign policy noted that the "United States is our premier partner in the Arctic and our goal is a more strategic engagement on Arctic issues." However, Plouffe continued, even as Harper attempted to align his government's regional foreign policy with that of the Americans, "the preferred strategy clashes with U.S. views." In particular, he noted that the prime minister's decision to relitigate the issue of the Northwest Passage, which had been sitting contentedly in a file marked "agree to disagree," caused irritation in policy circles in Washington, DC, prompting the country's Canadian ambassador to remark that "there's no need to create a problem that doesn't exist."

A brief attempt to remake Canada as a leader in the region was soon dashed on the rocks of realpolitik, a 2010 Arctic coastal states summit in Quebec to which only the United States, Russia, Denmark, and Norway were invited and from which Iceland, Finland, Sweden, and Indigenous nations were excluded, receiving something of a cold shoulder from other parties, including the United States.

Interestingly, wrote Plouffe, Canada's brief and atypical period of diplomatic robustness in the Arctic coincided with a turn by the United States, under Barack Obama, to a more inclusive and internationalist approach to the region—the kind of diplomacy more generally associated with Ottawa. By way of example, Plouffe cited Washington's three primary goals during its two-year chairmanship of the Arctic Council from 2015 to 2017 as "addressing the impacts of climate change in the Arctic, advocating stewardship of the Arctic Ocean, and improving economic and living conditions," with nary a mention of sovereignty to be found anywhere.

Of course, two years after Plouffe published his thesis, the pendulum swung again as Donald Trump swept into the White House with his own very particular view of global relationships. Perhaps predictably, the administration succeeded in ruffling the feathers of just about every participant in Arctic affairs, sometimes all at once.

At a meeting of the traditionally collegiate Arctic Council in 2019, Trump's secretary of state, Mike Pompeo, took swipes at the Arctic ambitions of China: "Beijing claims to be a near-Arctic state.... There are only Arctic states and non-Arctic states. No third category exists, and claiming otherwise entitles China to exactly nothing." And of Russia: "We're concerned about Russia's claim over the international waters of the Northern Sea Route," he said, before adding, "We recognize that Russia is not the only nation making illegitimate claims," a clear reference to Canada and the Northwest Passage. In case anyone was in any doubt, he pointed out that "the U.S. has a long-contested feud with Canada over sovereign claims through the Northwest Passage."

While none of the positions Pompeo articulated were necessarily novel ones, the nature of the discourse, as well as the setting of them, did not sit well. Michael Byers, a professor of political science at the University of British Columbia, told the *Guardian* that while Pompeo's remarks were consistent with US policy, the "belligerent" speech contained numerous "factual mistakes and logical inconsistencies." Gao Feng, China's special representative for the Arctic, sighed that the "business of the Arctic Council is cooperation, environmental protection, friendly consultation and the sharing and exchange of views. This is completely different now." And the Inuit Circumpolar Council (ICC) rebuked Pompeo, with Monica Ell-Kanayuk, then president of ICC Canada, stating that "the Northwest Passage is part of Inuit Nunangat, our Arctic homeland," and adding, "Mr. Pompeo's characterization of the Arctic as a place of geopolitical and military competition is faulty." She said, "Geopolitical differences in the Arctic have always been resolved peacefully. Indigenous peoples living in the Arctic are integral to its international institutions and decision-making that has achieved this."

The reason for Pompeo's bellicosity was clear. "Steady reductions in sea ice are opening new passageways and new opportunities for trade. This could potentially slash the time it takes [for ships] to travel between Asia and the

West by as much as 20 days," he said. However, the Trump administration refused to acknowledge the reason for those reductions, and the United States would not sign on to the Arctic Council's traditional end-of-session declaration because it opposed any mention of the words "climate change."

The administration was not quite done playing the role of skunk at the Arctic garden party, however. That same year, it confirmed that Trump had expressed interest in buying Greenland, a notion that Mette Frederiksen, prime minister of Denmark, which is responsible for Greenland's foreign and security affairs, dismissed as "absurd." Martin Lidegaard, a former Danish foreign affairs minister, called it a "grotesque proposal. . . . We are talking about real people, and you can't just sell Greenland like an old colonial power."

In a fit of pique, Trump promptly canceled a planned state visit to Denmark. Aleqa Hammond, the chair of Greenland's parliamentary foreign and security policy committee, said the effect of the whole episode was "at least one or two steps back" for the US reputation in the Arctic.

———————

There is no Northwest Passage. Strictly speaking, there are seven Northwest Passages.

In the hundred years or so since Roald Amundsen navigated the Passage from east to west, several different routes through the twists and turns of the Canadian Arctic Archipelago have been identified, each with its own advantages and disadvantages, depending on weather and ice conditions and the type and size of vessel attempting to make the journey.

Look at a map and the most obvious route will seem to be an almost straight line through what is now known as the Parry Channel and into the M'Clure Strait, a path that runs north of Baffin, Somerset, Prince of Wales, and Banks Islands and south of the Queen Elizabeth Islands. British explorer Sir William Edward Parry came close to traversing the Passage via this route in 1819–20 until blocked by ice; one reason that searchers failed to find the *Erebus* and *Terror* was an assumption that this was the path that Sir John Franklin would have taken.

It is the shortest, widest, and deepest route: the M'Clure Strait is just 170 miles long from east to west, 75 miles wide at its eastern entrance, and as much as a quarter of a mile deep. It is the passage favored by nuclear-powered submarines because of its lack of depth restrictions, but as relatively appetizing as it looks on paper, it remains one of the most hazardous because of ice that sweeps south into the M'Clure Strait from the Arctic Ocean. Old, thick, impenetrable ice is an almost omnipresent threat, and even under current ice loss scenarios it seems likely to remain so until deep into the twenty-first century.

A variant of this route, which follows the Parry Channel but then a takes a turn to the east and south of Banks Island via the Prince of Wales Strait, avoids the extreme ice in the M'Clure Strait but results in significantly diminished depth: the Prince of Wales Strait is just one hundred or so feet deep on average. It seems likely also to experience high ice levels through at least 2050. Nonetheless, this is the route that the SS *Manhattan* took on its westward journey.

There are a few variations to the most followed route, one of which the Franklin expedition was apparently attempting to navigate before its demise, one of which Amundsen did follow, and one of which the *St. Roch* used on its easterly voyage, all of which involve entering the Passage from the east via Lancaster Sound, then turning south to the east or west of Somerset Island, and then turning west to pass south of Victoria and Banks Islands. However, some stretches are difficult to navigate, include numerous poorly charted shoals and several ice choke points, and are unsuitable for ships with drafts of more than thirty feet.

The others are the domain only of much smaller vessels and, for depth and navigational reasons, unlikely ever to be of use as commercial passageways. Add to this the almost complete absence of infrastructure and the fact that, even during the limited navigational windows, which are presently primarily August and parts of September, sea ice remains a capricious presence—on our journey eastward out of the Passage, for example, we barely encountered any, whereas the voyage westward into the Passage barely two weeks before had required an icebreaker escort in places—the prospect of the Northwest Passage being a genuinely traversable and navigable alternative to the Panama and Suez Canals remains years, perhaps decades, in the future. And even then, the various twists, turns, narrows, shallows, and other challenges may work against the Northwest Passage being the passage that sees the most traffic in an ever-warming world. The Sturm und Drang of territorial tub-thumping is presently over a prospective route that very much lies in the future and may never match the vision of those who first set out to find a way to connect east and west through the Arctic.

But there may just be an alternative that might present itself more readily.

CHAPTER 6

An Arctic Bridge and the Polar Bear Capital of the World

In 1619, eight years after Henry Hudson sailed into the bay that now bears his name, two more ships, the *Unicorn* and *Lamprey*, followed suit, their crews weary from an unexpectedly taxing journey from Copenhagen. The Dano-Norwegian expedition, under the command of Jens Munk, was, like many of its time, searching for the Northwest Passage; but, having not arrived at the western edge of Hudson Bay until September, there was insufficient time remaining in the season to continue exploring, and so the crew steeled themselves for a winter ashore—the first Europeans to do so in the area.

They set up base near the mouth of what is now the Churchill River and initially found their surroundings pleasant and productive; there was an abundance of cranberries as well as fresh meat in the form of ptarmigans and caribou. There was also what must have seemed an alarmingly high number of polar bears, which prompted caution but also yielded its own supply of meat. But a productive fall gave way to a harsh winter, for which the Europeans were hopelessly ill-prepared, and of whom only three, including Munk, would survive. One by one they fell, not from the cold but from a mysterious disease that caused nausea, dysentery, and loss of appetite. The ground, hardened by winter, would not yield for the dead to be buried; even if it had, the survivors were too weak to carry out the task. Instead, corpses were dragged out onto the ice of the bay or left on the ground.

The return of spring brought with it renewed plant life, from which Munk and his two fellow survivors desperately sucked juices. As the ice on the bay broke up, they used nets to catch fish, which they boiled into a broth, their teeth too loose to enable them to eat solid food. By June, remarkably, they had gathered enough strength to leave, and they sailed the *Lamprey* out of the Churchill River, into Hudson Bay, and home.

One chilly November morning four centuries and change later, I am sitting in a truck near the site of Munk's encampment, where the Churchill River empties into Hudson Bay, and watching as a polar bear and her young cub wander along the ice that has formed near shore. The truck is owned by the province of Manitoba, and its driver, Ian Van Nest, is the conservation officer for the town of Churchill. For several weeks each year, his primary responsibility is keeping peace between the community's people and its resident polar bears, and he has spent much of the morning watching this pair of bears. His inclination had been to let them be and wander across the ice, happy to let them find their own way; but now, the mother's long, loping strides—and her cub's far shorter, more hurried ones—are bringing them closer to town, and for the safety of all involved, Van Nest determines that it is time to encourage them to change direction.

He exits the truck and asks assembled sightseers to retreat to the safety of their vehicles. Then he and a colleague take their shotguns and each shoots two cracker shells—twelve-gauge shotgun cartridges that are harmless but emit a loud bang when fired—into the air. Startled, the bears turn and run, soon slowing down and returning to their more leisurely pace as they stride out across the ice.

Inhabited by Cree and Dene tribes for over 1,500 years, Churchill and its environs were first settled by Europeans in the eighteenth century, when the Hudson's Bay Company established its northernmost fur-trading post there in 1717. Over the ensuing centuries, Churchill's fortunes ebbed and flowed; in the nineteenth century, as the fur trade declined, the community turned to whaling, targeting the bay's belugas for their oil. Following World War I, it was selected as the site for a new northern harbor, and a railway line was built to connect it to the provincial capital, Winnipeg, to the south. During World War II, the United States built a military base, Fort Churchill, a few miles to the town's east; when the war ended, the base became a joint US–Canadian facility for training and experimentation, from which more than 3,500 rockets were launched to study the ionosphere and the aurora borealis, which figures so prominently in the area's nighttime skies.

At the base's peak, more than 4,000 people lived in Churchill and the base. But in 1970, the United States left Fort Churchill, which was used only sporadically over the next couple of decades before shuttering completely in 1990. Today, the year-round population of Churchill is approximately 800; but each fall, the number of people in town swells as roughly 12,000 visitors and seasonal workers show up to take advantage of Churchill's latest incarnation: as the polar bear capital of the world.

The polar bears that live in Hudson Bay are the southernmost in the world, and as a consequence, they must contend with challenges that their more northern brethren do not always have to face: notably, each summer, the sea ice on Hudson Bay melts completely, and the bears must come ashore and wait until it forms again. They make temporary homes in earthen dens that have been hollowed out by generations of bears before them; they lower their metabolism, entering a state known as "carnivore lethargy," in which breathing is diminished and movement greatly reduced, until the cooling air of the fall prompts them to stir again, at which point they head for the shores of the bay and wait for it to freeze anew. In Western Hudson Bay, their path to the shore takes them close to, or even through, the town of Churchill.

Those were the bears that Munk and his crew saw; and it was those bears that killed almost all of them. The crew shot bears for food, not knowing that polar bear flesh is frequently infected with the roundworm that causes trichinosis. Most of the crew ate their meat parboiled in vinegar, which was insufficient to kill the parasite; Munk liked his bear meat more thoroughly roasted, while his two fellow survivors didn't care for it at all; their tastes proved to be their salvation.

Poisoning by eating polar bear meat is not a problem that afflicts Churchill's present-day inhabitants, but over the years the relationship between ursids and hominids has inevitably had its fractious periods. Coexistence was particularly fraught during Fort Churchill's heyday; the town's mayor, Mike Spence, told me that "it was common to shoot 25 bears in a season" back then. In the early 1980s, however, a loose collection of visiting photography enthusiasts and savvy residents realized that polar bears could be worth a lot more to Churchill alive than dead and began positioning the town as a tourist attraction. For tourists, Churchill provides a unique opportunity to see polar bears in the wild, which most do from onboard large buggies that look somewhat like school buses on monster truck wheels and that venture out onto the tundra as the bears laze on the shore of the bay. For those who reside in Churchill, living cheek by jowl with the world's largest living land carnivore makes for a unique experience and one that engenders a particular kind of caution. Residents rarely venture out on foot at night during bear season, and particular areas—along the banks of the river or the shores of the bay—are marked with warning signs. House and car doors are generally left unlocked in case anyone needs sudden shelter from a bear that may have walked around the corner.

While bad encounters can happen, they are rare: the last fatality as a result of a polar bear attack was in 1987, and while a young woman and her would-be rescuer were both badly mauled on Halloween night in 2013, nobody has been since. Signs around town provide a telephone number for people to call if they spy a bear in town; that number is answered by Van Nest, who with colleagues works to prevent polar bears and people from getting in each other's way. The first step is to do what he did the morning I was with him: give the bears the freedom to roam until they come too close to human habitation, at which point he aims to scare them off with cracker shells. Repeat offenders are caught in culvert traps and kept in what is officially known as the Polar Bear Holding Facility before being released onto the sea ice; but while caution and safety are watchwords, so too is respect for the species that was in Churchill first.

"Every time I deal with a bear, I think, 'Hey, thank you—thank you for gracing us with your presence, and it really was a pleasure meeting you,'" says Van Nest.

However, in recent years, the bears of Western Hudson Bay have faced a challenge far greater than the risk of running into a Churchillian. As temperatures have warmed, the sea ice of Hudson Bay has on average been forming later in the fall and breaking up earlier in summer, leaving less time for bears to hunt their favored seal prey. With energy and calories at a premium in the Arctic, that has manifested in a rapidly dwindling population: from an estimated 1,200 in 1987 to 935 in 2004 and just 618 in 2022. While some of that reduction may be due to some bears moving to the neighboring Southern Hudson Bay population, there are other signs of global warming's impact, including reductions in the body weight of adults and the number of cubs being born and surviving to maturity.

For polar bear researchers and environmental advocates, the concern that this engenders is obvious. For the people of Churchill, it is twofold: for the bears with which they have become synonymous but also for the town's economy. What is the Polar Bear Capital of the World without polar bears? Out of concern that Churchill may be reaching an inflection point and be close to needing another reinvention, thoughts are turning to a post–polar bear future, and in particular to the large port building that dominates the skyline to the northwest of town.

The first ship to bring cargo into the Port of Churchill was the British freighter *Pennyworth*, which arrived with a consignment of liquor, china and glass tableware, binding twine, and lubricating oil on August 17, 1932.

The port had opened one year previously after a six-year undertaking to connect Churchill and Hudson Bay to the rest of Canada by rail, and for decades it largely subsisted on domestic grain exports. In the early 1990s, however, the government of Manitoba began contemplating ways to expand its operations and to that end turned its gaze to the east—and, specifically, the Russian port of Murmansk.

In February 1993, Manitoba and Russia signed the Arctic Bridge Agreement, intended to identify and spur trade opportunities between Churchill and Murmansk across the North Atlantic. The Arctic Bridge would exit Churchill into Hudson Bay, pass through Hudson Strait into Davis Strait and Baffin Bay between Greenland and Canada, and travel south of Greenland and Iceland though the Norwegian and Barents Seas and thence to Murmansk. From either end, goods could be loaded onto rail and carried to points across Asia or south into Canada and the United States and beyond.

The plan has been, it is safe to say, slow to manifest. The 1993 announcement suggested the prospect of a pilot project that same year; by 2010, when Manitoba emphasized its commitment to the project, it was still no closer to coming into existence. Assessing the viability of an Arctic Bridge in a chapter in the 2013 volume *Shipping in Arctic Waters*, Willy Østreng noted that "for this bridge to materialize, Churchill will require significant additional port and rail investments, as well as further study by both countries regarding costs, cargos and volumes. The port of Murmansk is in need of similar improvements before this intra-regional traffic scheme—or intra-international route—can be realized."

A functional rail system is of particular importance at the Churchill end, given that there is no road into or out of town; and Churchill's vulnerability in that regard was never more starkly demonstrated than in 2017. Over the course of more than two days that March, an enormous blizzard dumped massive amounts of snow, which fierce winds drove into drifts that in places were twenty-five to thirty feet high. Large sections of road in town were buried under eight or nine feet of snow, and residents were forced to build tunnels to make it out of their homes. The response to the blizzard, local artist Sandra Cook told me, "brought out the best" in the community; but two months after the blizzard, Churchill faced a bigger challenge.

The snow melted.

The snowmelt produced flooding that washed out the final 155 miles or so of the railway tracks connecting Churchill to Winnipeg, and thus the world. The estimated repair bill was $43.5 million. OmniTRAX, the Colorado-based company that owned the track and which was already looking for a way to offload it, said it couldn't afford to foot the repairs and wouldn't. The federal government took the company to court. And Churchill found itself effectively isolated.

A day without the railway became a week. A week became a month. One month became eighteen. With no road connection to the outside world, the only way out for residents was a $1,200 round-trip air ticket, which few could afford. Everything that the town needed—every nail and screw, every can of beer, every pallet of produce—had to be flown in at exorbitant cost. Residents found themselves with far less cash on hand. Business at some stores in town dropped by 90 percent. Families scrimped and saved the money to fly out, never to return.

The beginning of the end of Churchill's near isolation came when a consortium of industry and First Nations communities banded together as the Arctic Gateway Group to buy the rail track and the port in August 2018. Repairs to the rail line began almost immediately, and on October 31 that year, the still of the Churchill night was pierced by the whistle of an approaching train.

In July 2019, ships resupplying Arctic communities returned to Churchill, and the port began shipping grain again that September. Grain shipments underwent a two-year hiatus while further repairs were made to the railway track, but Spence has stated that the intention is not only for those shipments to once again ramp up but also for the port to diversify. Churchill has reasserted itself as the primary resupply port for Arctic Canada, and there have been suggestions that it could also export crude oil from Alberta, although the prospect has divided the community. The port's backers have expressed enthusiasm about the prospect of it becoming a hub for the export of "critical minerals" for electric vehicles, batteries, and renewable energy. Manitoba's geology features twenty-nine of the thirty-one minerals listed as critical by the Canadian government, the industry has attracted increasing investment, and the province has drafted a minerals mining strategy to boost the industry while simultaneously addressing the somewhat conflicting goals of ensuring the minerals industry "conducts its activities in an environmentally sound manner," even as "environmental management and protection measures consider economic impacts."

There remain doubts about Churchill's viability as a major port. No matter how much the rail track is repaired and reinforced, it runs over miles and miles of muskeg—soft, swampy land that can cause tracks to buckle and warp and limits a train's speed. But if its infrastructure is presently lacking, it does at least exist, which puts it ahead of the Northwest Passage itself. The port may not give the impression of being ready for prime time, but it is the only deepwater port in Canada's north, and it does have four docks large enough for Panamax container ships. And, of course, shipping in and out is presently limited by ice conditions, with the port currently accessible only between late July and early November. But, at present, the Northwest Passage faces similar challenges; and Hudson Bay and environs are losing ice and experiencing longer ice-free seasons on average at a more rapid rate than points farther north in the Canadian Arctic. And there are fewer hazardous, ice-choked waterways to navigate in sailing out of Hudson Bay through Davis Strait and into the Atlantic than in traversing the Northwest Passage from west to east or vice versa. Churchill may face some competition in being Hudson Bay's port of choice: Manitoba is giving consideration to building a new port at Port Nelson, a ghost town that had been the originally intended terminus of the railway, to Churchill's southeast.

But it is a sign of the times that Churchill is planning for a near future in which it is simultaneously serving as home to the animal that is perhaps the most visible sign of climate change and positioning itself to take advantage of that climate change to survive. And while it doesn't by any means follow that an Arctic Bridge might be a full-blown alternative to shipping through the Northwest Passage, it could at least be complementary.

There is a tragic irony to the notion that, after all the lives that were lost in search of a Northwest Passage that could promote trade between east and west, the Passage could end up being beaten to the punch by a place that Europeans had first encountered in the 1600s and that had been home to a fur-trading post more than one hundred years before John Franklin set sail.

Somewhere, the ghost of Henry Hudson—banished to his death by his own crew for the crime of continuing his search for the Passage, and at least somewhat diminished in history for his failure to do so—might be permitted a wry smile.

———— ⌗ ————

On our final night in the Northwest Passage, word spread that the northern lights were visible in the sky above. *Ocean Endeavour* echoed to the sound of doors slamming and booted feet clomping along passageways and up stairs to the upper deck, where the aurora borealis revealed itself in its striking, silent glory. I found a quiet spot to sit by myself, drinking in the lights as they snaked and slithered through the night, fading in and out from view and reaching toward one horizon and then the other. I found myself reflecting on our journey, the places we had visited, the sights we had seen and the people we had met, and above all about the Passage itself. Here we were, fortunate enough to be on a comfortable ship in a temporarily ice-free passage, sailing east with nary a care in the world, all of us soon to be in the comfort of home, wherever that might be.

I thought back to being at the wreck of the *Erebus* and especially to something I had not been able to shake: the shoe that divers had brought to the surface shortly before we arrived there and to which the archaeologists on the barge were so tenderly caring. Who had worn it? How and when did he die? What would he have thought of the notion that, almost two hundred years after he set sail, the Northwest Passage would still exercise the thoughts of so many, or that he and his crew had played, even in death, a pivotal part in its history?

I thought, too, of Beechey Island and of having had the opportunity to pay tribute at the graves of those who died on that remote, desolate, windswept shore. They looked so lonely, if grave markers can: so far from home, so isolated, in such bleak surroundings. And yet they were lucky ones, their deaths sparing them from all that lay ahead, the frustration, the frightening realization that they and their ships were caught in the ice, the desperate and ultimately futile effort to strike out across King William Island.

When it was time to walk away, I kept looking back over my shoulder, taking one more photograph and then another, almost as if I couldn't bear to leave them alone again, until eventually I could see them no more.

Northeast

CHAPTER 7

"Our Island Is Falling Apart"

Little Diomede Island was in front of us, barely a stone's throw away, but we couldn't see it.

It was there, sure enough, its coastline clearly visible on the ship's radar. But even as we were almost upon it, we could detect no visible sign of it through the thick fog that engulfed everything around us. When, after several hours, the mist lifted, the island revealed itself in its stark, rugged splendor, its steep, boulder-strewn cliffs rising almost straight up into the clouds that persisted at its peak.

Two and a half miles to the west sat Big Diomede, 11 square miles in area, compared with the 2.8 square miles of its neighbor. Between the two lay the international date line—Saturday on Little Diomede is Sunday just a short distance away—and a border. Big Diomede is a part of the Chukotka Autonomous Okrug, Little Diomede an outpost of the state of Alaska, the two islands marking the closest point of contact between Russia and America. There are other differences: Big Diomede is uninhabited save for border guards and the staff of a weather station, denuded of its Native inhabitants by the Soviet Union in the 1950s to avoid repeated contact between those on either side of the imaginary line. On the base of Little Diomede, in contrast, lies a tiny village that clings to the small amount of exposed shore, whose eighty or so Inupiat Eskimo inhabitants survive primarily by fishing and hunting walruses, the ivory tusks of which they carve into ornaments and trinkets for trade and sale. The boundary between the two islands became known as the Ice Curtain; during our visit to Little Diomede in 1998, one resident told us that youngsters in the village would race across the water or the ice, aiming to touch Big Diomede and return without being fired upon by Russian border guards.

The islands are located in the Bering Strait, a narrow stretch of water that separates the Asian and North American continents and that at its narrowest point is a mere fifty-one miles wide. It is named after the Danish explorer Vitus Bering, who, under commission from Russia's Peter the Great, set out from Saint Petersburg in 1725 to sail from the Kamchatka Peninsula on Russia's east coast "along the shore that runs to the north and that (since its limits are unknown) seems to be part of the American coast." Having determined where it joined with the American continent, Bering was directed to find a European settlement, learn from the people there the name of the coast, make a landing, draw a chart with detailed information, and return home.

Bering, among the more cautious of explorers, did none of those things. He sailed up into the narrow strait that now bears his name, spied Big Diomede through binoculars, and sailed as far north as Cape Dezhnev, a promontory that is the easternmost point in Asia and the part of Russia that lies closest to America. Had he rounded the cape, he would have been able to ascertain that the assumption that the two continental landmasses were conjoined was wrong. Instead, wary of ice, he turned around and headed back home to present the fruit of his long labors: a partial coastline sketch of Kamchatka. Displeased, Peter enjoined Bering to try again; but by the time he left port once more, in 1741, he was deathly ill, and he died on what is now Ostrov Beringa, one of the Komandorskiye Ostrova, or Commander Islands, off the coast of Kamchatka in the Bering Sea to the strait's south.

In 1774, partly on the basis of Bering's reports, as well as those of Russian fur traders who hunted sea otters along the Aleutian Islands chain in the Bering Sea, a German historian and writer called Jacob von Staehlin published *An Account of the New Northern Archipelago*, which claimed the existence of a large island, located off the east coast of Siberia and separated from the North American mainland, called "Alashka." Four years later, British explorer James Cook, on a voyage to search for the Northwest Passage, investigated von Staehlin's assertions; he sailed north through the Bering Strait until being forced to turn around by advancing pack ice. Like Bering, he would not return home to tell the tale himself; he was killed a few months later in Hawaii following a conflict with islanders.

Indeed, the early history of European exploration of the strait was synonymous with disaster even before Bering. As early as 1648, approximately one hundred men in seven vessels under the leadership of Semyon Dezhnev and three others set out to explore the region, only for five of the boats to become lost or destroyed before reaching the strait. Dezhnev may have made landfall on the Diomede Islands and rounded the cape that now bears his name; but it would be almost twenty years before he made it back home, the only one of the four expedition leaders to survive. While on his return he was given two decades of back pay and earned a tidy sum from selling ivory from a walrus rookery he had encountered, his exploits were forgotten for a century and Bering was granted the glory of discovery.

Not that either man was even remotely the first to take to the water, or set foot on the land, around the strait. Even before the strait itself existed in the form in which we know it today, when water was locked up in glaciers and land levels were higher between 16,500 and 11,000 years ago, the area likely played a pivotal role in human history, when perhaps a few thousand people from eastern Siberia crossed what is now referred to as the Bering Land Bridge and became the first—or at least among the first—humans to settle in North America. Today, the Bering Strait region is occupied primarily by Inupiat, Siberian Yup'ik, and Yupi'it Eskimos, most of whom continue to depend on subsistence fishing and hunting of marine mammals. The strait is an important migration route for mammals and birds: gray and bowhead whales swim through it on their way to and from their Arctic feeding grounds, Arctic terns pause here during their globe-spanning flights between the northern and southern polar regions. Indeed, spring in the Bering Strait is a riot of activity as millions of birds and thousands of marine mammals follow the retreating sea ice north, pursued by polar bears and subsistence hunters.

In many ways, the environment has changed little over the last several millennia. In other ways, it has undergone profound transformations: gray and bowhead whales were hunted to near extirpation in the region, although grays have now rebounded to their pre-exploitation levels and bowheads are increasing in number. Just south of the strait, in the northern Bering Sea, warming has induced a shift from Arctic to subarctic conditions that has fundamentally changed the structure and composition of the marine ecosystem. Just north of the strait, in the village of Shishmaref, declining sea ice cover is exposing the coast to storm surges and causing the shore to erode at an average rate of up to ten feet per year. On Little Diomede, too, the impacts of a changing climate are being felt, not least in melting permafrost loosening the grip of the cliffs on the boulders that are strewn along their surface.

"Our island is falling apart," one resident, Anthony Soolook Jr., told me as we looked at it from the safety of our ship. "Last night I had a dream, that the boulders came down and smashed the village."

There is another change underway as well: incremental so far but potentially far greater in magnitude. The number of ship transits through the Bering Strait each year is increasing: from 262 in 2009 to 555 in 2021. Should warming maintain its trajectory and sea ice continue its retreat, those numbers seem likely to undergo further significant growth because the Bering Strait is the sole Pacific gateway to and from both the Northwest and Northeast Passages.

————— ∞ —————

To sail north through the Bering Strait, you must first sail north through the Bering Sea. And to enter the Bering Sea, you must pass through the Aleutians, a chain of fourteen large volcanic islands and fifty-five smaller ones that reaches out from the southwestern point of mainland Alaska and curves across the northernmost reaches of the Pacific Ocean like a string of pearls. The chain ends with the Russian-owned Commander Islands, on which Vitus Bering met his demise, but the bulk of them are Alaskan, and because the westernmost of those, Semisopochnoi Island, is on the western side of the 180-degree longitudinal meridian, Alaska is effectively both the westernmost and most easterly state in the union.

For at least four thousand years, the archipelago has been inhabited by the Aleut people, who refer to themselves, depending on from which end of the archipelago they hail, as Unanga or Unangan; they numbered an estimated 25,000 when Russian fur hunters first arrived on the islands in Bering's wake, before the population plummeted to just a couple of thousand as a result of violence inflicted and diseases introduced by the invaders. Today, close to 12,000 identify as being Aleut, while 17,000 or so claim at least partial Aleut ancestry.

The Aleuts were even caught in the crossfire of World War II. The 45 Native inhabitants of Attu Island, in the chain's far west, were captured as prisoners of war and sent to Hokkaido when Japanese forces invaded and occupied Attu and neighboring Kiska Island in 1942; the United States government evacuated much of the rest of the island chain and sent 880 Aleuts to internment camps in southeastern Alaska, where almost 100 of them died.

One of the Aleutian Islands, Amchitka, was selected by the US government as the location for a trio of atomic bomb tests from 1965 to 1971, the last of which was the largest underground test ever conducted by the United States. Concerns that the shock waves from the detonation would send a tsunami toward the Pacific coast of North America prompted a group of Quakers in Vancouver to hatch a plan to sail a ship to the test site; calling themselves the Don't Make a Wave Committee, they signed off one planning meeting with the word "peace," prompting one of them to add that they should make it a "green peace." And thus was one of the world's largest environmental organizations born.

That first Greenpeace voyage never did reach its target, foiled by the attentions of the US Coast Guard; not until 2009 did a Greenpeace ship finally arrive at Amchitka, by now a nature reserve, with me onboard, as part of a broader investigation into the ecosystem of the Bering Sea region. We scrambled across the kelp-covered rocks ringing the island, made our way past the leftover military materiel, and stood on the shores of the lake that had been formed by the 1971 blast, looking off into the distance and, in a silence interrupted only by the rustling of grass in the wind and the occasional chirp of a passing bird, contemplated the violence that had been visited on this remote outpost.

In addition to the prospect of a tsunami, one of the strong motivators for the protest was the possible impact of the blast on the region's sea otters—of which, during our visit, we saw hardly any. Having long since been greatly depleted by fur hunters, the sea otters of the Aleutians have now further fallen victim to the profound changes in the Bering Sea ecosystem, their continued decline partly fueled by orcas, which had previously paid little attention to the aquatic mustelids, turning to them for nutrition when their traditional prey of whales and sea lions became less numerous as a result of a combination of human predations and the disruption caused by overfishing and warming waters. As sea otters diminished in number, the urchins and abalones they ate in vast numbers—a single sea otter being capable of consuming as many as one thousand urchins per day—multiplied rapidly. Unchecked, they laid waste to the area's kelp forests and then turned their attention to the coralline algae that blanketed the cold-water reefs on which those forests once stood. Something similar had happened in the past, of course, when fur traders hunted the region's otters to near extirpation; one difference this time is that warmer waters are more acidic waters, and the greater acidity has chipped away at the resistance of the algae's reef skeletons. At the same time, the extra warmth increases animals' metabolisms, making the urchins even more voracious. And ultimately, of course, when the sea otters declined in the past, the fur hunters moved on, allowing the otter population to rebound and the ecosystem to recover. This time around, the pressures are, if anything, only increasing, as a warming world is all too widely seen as at worst an inconvenience to be endured and at best an opportunity to be exploited, rather than what it truly is: a planetary experiment the impacts of which are even now only beginning to be felt.

The principal route through the Aleutians, even for those who are arriving from the islands' southwest, is near the eastern end of the island chain, a nine-mile-wide strait known as Unimak Pass, which lies west of an eponymous island and the imposing, conical Mount Shishaldin, one of the ten most active volcanoes in the world.

The cold, nutrient-rich waters that flow through Unimak Pass carry a wealth of plankton into the Bering Sea, providing a platform on which is built the bounty of the region's ecosystem and its fisheries, still profitable even in a time of ecological flux: according to the Alaska Seafood Marketing Institute, Bering Sea and Aleutian Islands fisheries generated a wholesale value of $2.66 billion in 2019 alone. That constituted 58 percent of the amount generated by all of Alaska's fisheries and does not account for the billions of dollars generated by fisheries on the Russian side of the Bering Sea; but the ecological and economic value of the region does not mean that it is spared from other human activities.

In 2012, Admiral Robert J. Papp Jr., now retired but then commandant of the US Coast Guard and subsequently the US Department of State's special representative for the Arctic, noted of Unimak Pass that "there are literally thousands of ships that transit through there, carrying fuel and other things that were at risk for environmental disasters, sinkings, and other things." Those thousands of vessels, far higher than the number that travel farther north to traverse the Bering Strait, include cargo and container ships moving between the Pacific Northwest and British Columbia on one side and China on the other, which often must contend with the strait's frequently severe weather and strong tidal flows. In 2004, *Selendang Ayu*, a Malaysian cargo ship, had just cleared the pass when it lost power in a storm, was blown aground, and broke apart, spilling 1.2 million liters (7,500 barrels) of fuel oil—virtually none of which, because of the accident's remote location and the lack of search and rescue facilities in the Aleutians, was recovered.

There are no restrictions on traffic passing through Unimak Pass because the United States—showing, at least, an admirable consistency in such matters—views it as an international strait, as a result of which there are no notification or pilotage requirements and no designated shipping lanes. Until recently, there were no shipping lanes in the Bering Sea, either, but in 2019, the International Maritime Organization adopted a joint US–Russian proposal for a series of two-way shipping routes in the sea and through the

Bering Strait. The lanes, which are voluntary—but which, according to a 2020 analysis by The Pew Charitable Trusts, most large ships soon began to follow—are recommended primarily for vessels over four hundred gross tons and include three environmentally and culturally sensitive areas to avoid.

Heading north from the Aleutians, the first such area surrounds Nunivak Island and the approximately two hundred Yupi'it Eskimos who live in the island's one settlement, the city of Mekoryuk. The principal shipping lane lies to the west of Nunivak and continues north, passing east of the second area to be avoided, the waters surrounding Saint Lawrence Island, reckoned to be one of the last remaining exposed areas of the Bering Land Bridge and home to the subsistence whaling communities of Gambell and Savoonga as well as to an abundance of marine mammals and seabirds nurtured by the cold, nutrient-rich Anadyr Current. Continuing north to the entrance to the Bering Strait, the lane skirts the waters of tiny King Island, now largely uninhabited but the source of a distinct Eskimo culture; although most now live on the Alaskan mainland, many King Islanders return to hunt walruses and seals.

Prior to reaching King Island, the lane passes west of Alaska's Seward Peninsula, on the south coast of which lies the town that was once the most populous in Alaska and that in more recent times has been pitched as the possible site for a deepwater port to service ships coming to and from the Northwest and Northeast Passages. Incorporated in 1901, Nome and its environs first came to the attention of the Western world three years earlier, when a trio of prospectors discovered gold. Within a year, the area's non-native population soared from zero to 10,000, its numbers including the likes of Wyatt Earp, who headed north in search of riches after an eventful life that included being a peace officer in Kansas; playing a key role in the gunfight at the O.K. Corral in Tombstone, Arizona; acting as head of security for William Randolph Hearst; and even serving as referee in a world heavyweight title fight. Within a decade, however, the gold rush had run much of its course and Nome's population had fallen to 2,600; in 1925, it succumbed to an epidemic of diphtheria that was broken by a relay of dog sleds bringing a supply of antitoxin serum that saved the town and is today commemorated by the annual Iditarod Trail Sled Dog Race.

Across the water from Nome, sheltered in a deep fjord on the Russian coast, Provideniya is a quiet port town of perhaps two thousand people, described by one writer as being "absolutely beautiful . . . clinging to the narrow fringe of Komsomolskaya Bay and surrounded by towering stark mountains cloaked with swirling low cloud. . . . Even in a treeless tundra environment, Provideniya and its surrounds were nothing short of breathtaking." And yet, Provideniya "was a paradox; its exquisitely beautiful setting scarred by an endless rubbish heap of abandoned villages, corroding Soviet buildings and old stone military barracks." Like Nome, Provideniya once had been a larger, more bustling presence, a town of more than nine thousand that experienced a rapid decay following the collapse of the Soviet Union. Developed largely to be the port that marked, depending on your direction of travel, the entry point into or final stop on the way out of the Northeast Passage, it retains that role today and nurtures the prospect of returning to its former glories as vessel traffic increases.

From just outside Provideniya, another shipping lane heads northeast toward the Diomede Islands. Shortly after passing King Island, the lane that heads north from the Aleutians splits into two, one branch steering west of the Diomedes, intersecting with the lane from Provideniya, the other continuing north into the Bering Strait, east of Little Diomede, past the Alaska township of Wales—where a sculpture of an outstretched hand releasing a dove flying toward Siberia is matched on the Russian side by an *umiaq* preparing to make the journey of friendship across the strait—and Shishmaref (at which point the suggested shipping lanes end) and into the Chukchi Sea. From there, any ship continuing northward can, if it is so inclined, head past Point Hope in Alaska's northwest, round the corner, and head east into the Beaufort Sea. Continuing east, Alaska's northern coast passes by on the starboard side, from the village of Wainwright to the larger town of Utqiagvik, the most northerly settlement in the United States, and onward past the oil fields of Prudhoe Bay and the village of Kaktovik, across the border, and into the morass of waterways of the Canadian Arctic Archipelago.

Alternatively, returning to the point west of Big Diomede, you can steam north to Cape Dezhnev and the village of Uelen, site of a settlement that for the best part of two thousand years subsisted largely on fishing and the hunting of walruses and is now home to the world's only museum of walrus ivory carvings.

Continue around Cape Dezhnev and to the north and west and soon our hypothetical vessel encounters the twin sentinels of Wrangel Island and its neighbor, Herald. Surrounded by an expanse of ice-covered sea, the islands present imposing but contrasting vistas: the larger Wrangel, at roughly ninety by fifty miles approximately the size of Crete, boasts northern and southern coastal plains that rise gently to meet a ridge of mountains, on average 1,600 feet above sea level, that stretch along the middle of the island from east to west like a backbone; and the much smaller Herald, a mere four and a half square miles in area, rising directly upward into jagged cliffs. When our ship approached the islands in 1998, shortly after we visited the Diomedes, our bosun, worldly wise and slow to be impressed, gazed at Herald's cliffs disappearing into a low layer of cloud and declared it to be like "something out of Arthur Conan Doyle's *The Lost World.*"

Wrangel and Herald seem far removed from human concerns, but one hundred years ago, Wrangel was at the center of a diplomatic storm and a pair of disastrous expeditions, both engineered by Vilhjalmur Stefansson. Seven years after the disaster of the *Karluk*, Stefansson sent five settlers to live on the island as a means of establishing occupancy and sovereignty for Canada; the only survivor was Ada Blackjack, an Inupiat seamstress who was brought onboard in Nome. Canada distanced itself from the whole affair, and when Stefansson sent a further group of settlers after Blackjack was eventually rescued, they were soon removed by Russia.

Wrangel Island was the last redoubt of the woolly mammoth, the final representatives of which expired approximately four thousand years ago, five millennia after the species had all but disappeared from the Arctic mainland. It remains home to major breeding populations of walruses, bearded and ringed seals, Arctic foxes, wolverines, and wolves, and it boasts the highest density of polar bear denning sites in the world. As our small icebreaker threaded its way through ice floes that appeared to come at us in waves, as if protecting the two islands, frequent sightings of polar bears swaggering across the floes repeatedly prompted a rush of crew onto the deck to watch them as they, in turn, nonchalantly monitored our progress through their realm.

Our journey north had been notable for its absence of sea ice, thwarting the collective goal of the scientists onboard who were keen to study life at the ice's edge. On our approach to Wrangel and Herald, the ice appeared en masse as if out of nowhere, all but daring us to attempt further progress. Arne Sorensen, the experienced Danish captain at the helm of our small icebreaker, nudged the smaller floes out of the way; the larger ones he approached directly, slowing down before impact like a charging bull with second thoughts. The ship would rise onto the ice and bear down on it, either opening a crack that Sorensen could work into a full split or shoving the floe to one side. The captain's task was not aided by the descent of a thick layer of fog that prevented him from anticipating the scene ahead and forced him to react to each new floe or chunk of ice that emerged toward us out of the murk.

Eventually, the fog cleared and the ice relented. Wrangel and Herald lay directly ahead; beyond them stretched the almost three thousand miles of water and ice that constitute the Northeast Passage.

CHAPTER 8

The Rise and Fall of Mangazeya

Murmansk is the largest city north of the Arctic Circle and one of the largest ports in Russia. Stretching 12 miles along the mountainous coasts of Kola Bay, it is at once remote and accessible: more than 1,200 miles north of Moscow, it is a mere 65 miles from the Norwegian border and a roughly ninety-minute flight from Helsinki. It is both grand and imposing—its main streets are wide and open, with streetcars running regularly along their length—and strangely provincial: when our charter flight touched down from the Finnish capital, we were shuffled into a small room, where we milled around for a disorganized passport check and luggage retrieval before spilling out directly into the street. Its boundaries contain both grandeur and squalor in a way that is not uncommon in large cities anywhere but that in some respects also seems a very specifically Soviet hangover. Statues and monuments, once gleaming, are, upon close inspection, often coated in coal dust, a consequence of open-air coal transshipments to Europe via the Murmansk Commercial Seaport.

Geographically, economically, and culturally, the seaport is the heart of Murmansk. It is the reason the city was constructed in 1915, as the coastal terminus for a newly built railway constructed to transport military supplies from allies into the heart of Russia during the early days of World War I. It was originally dubbed simply the Murman Terminus, after the Murman Coast on which it is located, the name of which itself may be a corruption of an Old Norse word for "Northman." In September 1916, having grown rapidly, it was formally granted cityhood and dubbed Romanov-on-Murman, after the Imperial Russian ruling family; following their deposition the following February, it was swiftly renamed Murmansk.

The city's connection to the Arctic and the ocean is evident throughout. The *Lenin*, launched in 1957 as the first icebreaker—as well as the first surface ship and civilian vessel—to be nuclear powered, is berthed here permanently as a museum. The deckhouse of the nuclear submarine *Kursk*, which was based in Murmansk but which sank to the bottom of the Barents

Sea in 2000 with the loss of all 118 personnel onboard, has been retrieved from the ocean floor and repurposed as the centerpiece of a memorial to sailors who died in peacetime. High above the water, looking down on the ships arrayed along the docks, stands an imposing statute of a solider in a greatcoat, submachine gun slung over his shoulder, known informally as Alyosha and formally as the monument to the Defenders of the Soviet Arctic during the Great Patriotic War. Standing 116 feet tall atop a 23-foot-tall pedestal, it looks west toward the Valley of Glory, site of some of the fiercest fighting to take place in the Russian Arctic during World War II.

Despite Murmansk's northerly location and subarctic climate, its waters are ice-free year-round, courtesy of the warm North Atlantic current that reaches all the way up into the waters of Kola Bay, and as a result it is the main hub for Russia's Barents Sea fishing fleet. It is also the home port for the entirety of Russia's nuclear-powered naval fleet and is the putative eastern terminus of the Arctic Bridge from Churchill. Prior to Russia's invasion of Ukraine, the port turned over sixty million tons of commercial cargo each year. The key port in the Russian Arctic, Murmansk also fashions itself as the "headquarters" of the Northeast Passage.

Had Murmansk existed in 1878, perhaps Adolf Erik Nordenskiöld might have paid a visit onboard *Vega*—a three-hundred-ton vessel built for walrus hunting and with a hull that was strengthened against incursions from ice. But his journey would end up playing a significant role in establishing that the Northeast Passage was at least somewhat navigable, which in time would become a major element of Murmansk's raison d'être.

Born in 1832 in Finland, which was then under Russian control, Nordenskiöld moved to Sweden twenty-five years later when his public agitation against Saint Petersburg's rule attracted uncomfortable amounts of attention from the authorities. There, he became curator of mineralogy at the Swedish Museum of Natural History and developed a keen interest in Arctic exploration. In 1861, he participated in an expedition that attempted to reach the North Pole with dogsleds across the sea ice from the north coast of Spitsbergen. In 1868, he sailed farther north than anyone had done previously, in 1870 he explored western Greenland, and in 1872 he again embarked on an expedition to Svalbard—as the archipelago is now known, with Spitsbergen relegated to the name of the largest island—this time with the intent to reach the Pole with sledges drawn by reindeer.

His polar ambitions thwarted by thick ice, he shifted his attention else-where, and after a pair of expeditions in 1875 and 1876 reached the Kara Sea and confirmed the feasibility of establishing a trade route between Europe and Siberia, he resolved to return with a goal to explore the little-known area to the east and navigate the Northeast Passage.

His plan, he wrote, was to "leave Sweden in July 1878 in a steamer specially built for navigation among ice, which will be provisioned for two years at most." He would set course for the Russian Arctic archipelago of Novaya Zemlya, where he would wait for the ice to yield enough for him to enter the Kara Sea. From there, he would continue to Cape Chelyuskin, the most northerly point on the Asiatic continent, "where the expedition will reach the only part of the proposed route which has not been traversed by some small vessel, and is rightly considered as that which it will be most difficult for a vessel to double during the whole North-East Passage; but our vessel, equipped with all modern appliances, ought not to find insuperable difficul-ties in doubling this point, and if that can be accomplished, we will probably have pretty open water towards Behring's Straits, which ought to be reached before the end of September."

He was able to execute the first half of the plan with some alacrity, reach-ing Cape Chelyuskin on August 19 and, as the fog lifted, spying a polar bear. He had, he wrote, "reached the great goal, which for centuries had been the object of unsuccessful struggles. For the first time a vessel lay at anchor off the northernmost cape of the Old World. With colors flying on every mast and saluting the venerable north point of the Old World with the Swedish salute of five guns, we came to anchor."

The hardest work lay ahead, however. Fog and ice bedeviled their progress over the next several weeks, with *Vega* able to continue only by hugging a narrow ice-free path close to shore. In time, however, even that route became unavailable, and on September 28, the *Vega* was held prisoner by the ice. They had traveled in excess of 2,000 miles and lay a mere 120 miles from the end of their journey, but they had no option but to sit and wait.

Fortunately, their entombment was close to a settlement of Chukchi natives, with whom the crew of the *Vega* struck up a relationship over the course of the winter and beyond. Although the Chukchi did not speak Rus-sian, they did know a few words of English, courtesy of visits by American whalers who had traveled up the Bering Strait and turned west; in time the two camps were able to enjoy rudimentary conversations, and the unex-pected company appears to have boosted morale on both sides.

Not until the following June, after several false dawns in the late spring, was the *Vega* finally able to break free and complete its voyage. Nordenskiöld turned south into the Bering Strait and, on August 14, anchored at the Russian settlement on Ostrov Beringa. And then he kept going: to Japan, where *Vega* underwent repairs; north of Indonesia and into the Indian Ocean; south of India and through the Suez Canal; into the Mediterranean and out via the Strait of Gibraltar; north via the English Channel and through the North Sea, before eventually returning to Stockholm, amid much fanfare, on April 24, 1880, having become the first to circumnavigate the Eurasian continent.

Nordenskiöld's voyage is considered one of the finest achievements in Swedish exploration and, indeed, in polar exploration more generally. He himself was widely feted for his achievement—receiving a medal from the emperor of Japan and being awarded a baronetcy by King Oscar II. He had established that the Northeast Passage is navigable, if only with some difficulty.

But while the Northeast Passage includes the waters of the Barents Sea north of Scandinavia, fully 90 percent of its extent lies farther east, beyond Novaya Zemlya and north of the coastline of just one nation. It was that nation that picked up Nordenskiöld's baton and not only has committed itself to making the passage a functional waterway but also, particularly in the latter years of the twentieth century and throughout the twenty-first, has presented it as an integral part of its territory, its economy, and its very being.

In an essay in a 1991 publication titled *The Soviet Maritime Arctic*, William Barr of the University of Saskatchewan wrote that the "most compelling aspect of Russian involvement in the Arctic Ocean and its peripheral seas is that it spans five centuries." As early as 1478, he noted—almost a century before Martin Frobisher set out in search of the Northwest Passage, and the best part of eight decades before what would become the Muscovy Company dispatched three ships eastward toward the Northeast Passage—Ivan III conquered Novgorod, to Moscow's northwest, in the process acquiring territory that reached to the coast of the White and Barents Seas and giving Russia its first access to a seacoast. By the middle of the sixteenth century, Russian craft were sailing as far east as Ostrov Vaygach, just south of Novaya Zemlya, to catch and hunt walruses. And the early years of the seventeenth century saw the development of a coastal route known as the Mangazeya Sea Route, which extended east from the mouth of the Northern Dvina River, near present-day Arkhangel'sk (roughly 370 miles southeast of Murmansk), along the coast and via portage across the Yamal Peninsula, to the mouth of the Taz River and the Siberian city of Mangazeya. (The sheer scale of Russia is sometimes hard to comprehend; to get a sense of its immensity, consider that Mangazeya was on the approximate longitude of eastern Pakistan and Kyrgyzstan, which is merely halfway or so from Murmansk to Cape Chelyuskin, which is itself a very approximate halfway point along the Northeast Passage.)

Founded by Cossacks in 1600, Mangazeya was a trading post for fur and walrus ivory, which was then shipped west along the coast to Norwegian, English, and Dutch merchants. Within a short space of time, it had grown into a "fabulous polar city" and a "virtual Baghdad of Siberia," from which "hundreds of thousands of sable, ermine, silver and blue fox skins, and countless tons of precious mammoth and walrus ivory" were exported to Europe. However, there was a catch. Mangazeya's trade sprang up during a time in Russian history known as the Time of Troubles, which began in 1598 when Tsar Fyodor I died without an heir and concluded in 1613 with the accession of Mikhail Fedorovich; in the interim, six claimants to the throne came and went, Moscow was occupied by Polish forces that took advantage of Russia's internal distractions, and a famine from 1601 to 1603 claimed the lives of one-third of Russia's population. In the midst of this turmoil, trade along the Mangazeya Sea Route was able to take place without interference or oversight; once Mikhail assumed power and normalcy began to return, however, its freewheeling days were numbered.

Mangazeya's booming trade aggravated inland merchants, who saw it as taking away business that would otherwise have come their way. They agitated for the city's unfettered activities to be wheeled in: Mikhail, meanwhile, looked at the Mangazeya Sea Route and saw not just a means by which merchants could operate without control and taxation, but also a pathway via which English merchants under the direction of the Muscovy Company could gain unfettered access to all of Siberia. He even feared that it was a back door through which England could seek to assume control of Siberia and claim it as a colonial bauble.

And so, in 1619, Mikhail shut it all down. He closed the sea route completely, even to Russians, for fear that they might divulge its location to outsiders, and declared that anyone who disobeyed should be "put to the hardest possible death, and all their homes and families destroyed branch and root." According to Benson Bobrick in his book *East of the Sun: The Epic*

Conquest and Tragic History of Siberia, "navigational markings were torn up, surveillance posts established along the coast to intercept and kill all those who attempted to get through. . . . Maps were even falsified to depict Novaya Zemlya as a peninsula rather than an island at some cost to later mariners who would rely upon them as nautical guides."

Mangazeya subsequently declined and, in 1678, was burned to the ground. In time, it came to be forgotten entirely for the best part of three hundred years: the first true trading post along the Northeast Passage, and the first to fall victim to the geopolitics of the Northern Sea Route.

———∞———

Over the course of the seventeenth century, Russian explorers and traders made their way up Siberia's river systems, establishing outposts along the way, until they reached the Arctic coast, at which point they edged their way westward through ice floes and eastward along the coast. Their vessels were initially of primitive construction, made almost entirely of wood with a deer hide sail, although they became progressively more robust, reaching up to sixty feet in length; constructed from pine or larch; braced with crossbeams; reinforced with iron nails, spikes, and bolts; and fitted with a keel and canvas sails.

Known as *koches,* these were the boats that Semyon Dezhnev and comrades used to explore the Bering Strait region; but with their efforts becoming overlooked and forgotten, the idea of establishing a true trade route along the length and breadth of Russia's Arctic coast seemed less and less like one to be taken seriously. The area, said one official, "is utterly unnavigable with Ships, and should a second Christopher Columbus appear, and point out the course of the heavens, he could not yet drive away these Mountains of Ice: For God and Nature have so invincibly fenced the sea side of Siberia with Ice, that no Ship can come to the River Jenisea, much less can they come farther Northwards into the Sea."

After Peter the Great came to power in 1683, however, he was intrigued by the notion of a Northeast Passage that, he was advised, could not only theoretically shave the sailing time between Europe and Japan from nine months to five or six weeks, but could also be a source of income if European powers were charged a toll to use it. That was the impetus for his commissioning Vitus Bering's voyages, but their underwhelming results again caused most in Russia to turn away from thoughts of a viable sea route along the Arctic coast.

The success of Nordenskiöld and other foreign explorers brought a brief resurgence of Russian interest in the late nineteenth century, notably with the commissioning of the first polar icebreaker, the *Yermak*, in 1898. Its construction was authorized by Admiral Stepan Osipovich Makarov, who intended it for Arctic research and development of the Northern Sea Route, but he was strongly opposed by bureaucrats in the finance and navy departments. Undeterred, he commanded the *Yermak* on its maiden voyage to Svalbard, but after it sustained hull damage in the ice, it was consigned to duty in the Baltic Sea.

The first two decades of the twentieth century, however, saw Russian attempts to develop the Northern Sea Route ramp up significantly in response to a series of external pressures brought about by the country's turbulent start to the 1900s. The 1904–5 Russo-Japanese War placed immense pressure on the Trans-Siberian Railway and provided the stimulus for developing the Northern Sea Route, as did Russia's defeat in that conflict. From 1910 to 1915, the Arctic Ocean Hydrographic Expedition saw two small icebreakers, *Taymyr* and *Vaygach*, steam from western Russia to the Pacific via the Suez Canal and Singapore, and then spend three years slowly working their way westward through the passage, sounding and surveying as they went and returning each winter to the far eastern city of Vladivostok.

The tumult of World War I and the Russian Revolution added further impetus to development of an Arctic sea route. In 1920, the now Soviet government established the Komitet Severnogo Morskogo Puti, abbreviated to Komseveroput, or Committee of the Northern Sea Route, to "equip, improve, and study" the route from Arkhangel'sk to the Bering Strait. Twelve years later, Otto Schmidt completed the Northern Sea Route (NSR) in the Soviet icebreaker *Sibiryakov*—more or less: the ship lost a propeller in the ice just shy of the Bering Strait and had to be towed the rest of the way by a freighter after emerging from the ice under improvised sails. That same year saw the formation of a new and more powerful government department, the Glavnoye Upravleniye Severnogo Morskogo Puti (Glavsevmorput), or Chief Directorate of the Northern Sea Route, with Schmidt as its head.

Over the next decade, traffic along the Northern Sea Route increased substantially. In 1934, *Fedor Litke* became the first ship to transit the entire route in one season without suffering an accident. The following year, it accompanied the first freighters, *Vantsetti* and *Iskra*, through the route from west to east, and in 1936 it escorted the first two warships, *Voykov* and

Stalin, along the route from east to west; that same year, twelve freighters made the passage from west to east and two from east to west.

By the outbreak of World War II, Soviet authorities had "come close to fulfilling their intent of making the Northern Sea Route a regularly operating transport artery." It remained, however, a primarily domestic operation; Peter the Great's dream of making it a passageway that other countries would willingly pay to use remained unfulfilled for many years.

In 1940, before the Soviet Union entered World War II, the *Komet*, a Nazi raider disguised as a merchant ship, was escorted along much of the NSR in return for a five-figure fee because Adolf Hitler wanted to sneak a patrol ship into the Pacific. At some point, however, Soviet authorities appeared to have a change of heart; about eight hundred miles short of the Bering Strait, they advised the *Komet* to turn around and head west, but the German captain refused and continued the rest of the way unescorted. Whatever Moscow ultimately felt about its role in this particular endeavor, it did demonstrate, as one analysis put it, that "the Soviet Union was capable of moving manpower and equipment between the Atlantic and the Pacific Oceans without leaving its territorial waters."

The *Komet* was the first foreign-flagged ship in over twenty years to be granted passage through the NSR, and it would be a further fifty years before the next one. A 1967 offer to provide icebreaker escort to foreign ships was never acted upon and may have been quietly withdrawn to avoid offending Moscow's Arab allies, who would not have taken kindly to the country pushing an alternative to the Suez Canal. Only following a 1987 speech by Mikhail Gorbachev, which we'll look at in more depth later and which essentially offered an open invitation to other nations to revisit that 1967 offer, did the Northern Sea Route experience its first real stirrings of international trade. The Soviet Union offered shipping companies the ability to lease cargo space onboard SA-15 icebreaking cargo carriers; in 1993, one such company chartered an SA-15 to ship timber from Helsinki to several ports in Japan via the NSR.

One company executive stated: "We avoided the $90,000 cost of using the Suez Canal and reduced our transit time by 14 days with ship costs that run about $10,000 a day. Based on our experience, I couldn't imagine it not becoming a major international shipping route."

CHAPTER 9

Red Arctic

Of the 195 countries in the world, 8 can be considered truly Arctic: the United States, Canada, Russia, Norway, Denmark, Iceland, Sweden, and Finland. These countries constitute, along with (at time of writing) 13 observer countries and 6 Indigenous representative organizations, the membership of the Arctic Council, which was formed in 1996 to address issues faced by the nations, inhabitants, and environment of the region. The 8 Arctic countries have equal votes in the forum, the 13 observers have none; even so, in the Arctic as in other walks of life, some countries are in practical terms more equal than others. Some carry the swagger of a superpower; others engage in more activities in the Arctic, with a greater capacity to impact the region; and others are simply more invested: economically, politically, environmentally, or even psychologically.

In terms of economic and military might, the United States remains, for now, without peer. It has, however, traditionally been less significant a player in the Arctic than one might imagine given its general geopolitical heft, largely because the period between nineteenth-century American-led explorations in search of the Open Polar Sea and a twenty-first-century realization that it needed to engage seriously in the region was characterized by what might be termed a studied indifference. To a large extent, among the broader body politic and the general populace, that remains the case: a 2016 survey of American citizens, for example, found that just 18 percent of those polled knew the United States is an Arctic nation, with many respondents convinced it "was a trick question to which we were offering no correct answer." Perhaps this isn't entirely surprising, given Americans' famously subpar geographic knowledge and the fact that all of the country's Arctic seas and lands are in a state that joined the union only in 1959, does not border any other state, and is all too frequently left off maps of the United States or relegated to a small corner. And while the Pentagon has of late been sounding the alarm over the security challenges of a melting Arctic, by and large American policy toward the region has been subcontracted to the Alaska congressional delegation—which in recent years effectively means the state's senior senator, Lisa Murkowski, whose diligent statecraft and policy chops have earned her a reputation as something of an éminence grise among political figures in the region.

In contrast, Canada, rarely considered a geopolitical heavyweight, is something of an indispensable nation in the Arctic—even if it almost stumbled into Arctic governance when Britain dropped responsibility for its northern regions into its lap. Its opinions and positions would carry weight anyway, given that it boasts more than 100,000 miles of Arctic coastline and that the Canadian Arctic Archipelago alone encompasses an area of some 500,000 square miles. Canada, more than most, is able to speak with some authority and experience on the issue of Indigenous rights and perspectives, having undergone a relationship with the Arctic's founding peoples that, as we have seen, has at times involved some horrendous misjudgments, mistakes, and outright racism but which has also, in recent decades, been undergoing a significant course correction—not yet by any means sufficient, perhaps, to compensate for generations of deep and profound mistreatment, but a substantial start along the lengthy road to redemption. And, of course, Canada is inevitably at the center of discourse over the Arctic's future given the contentiousness and uncertainty that surrounds the fate and status of the Northwest Passage.

And then there's Russia. Its northern shore, stretching from the Barents Sea to the Sea of Okhotsk in the east, accounts for 53 percent of the Arctic Ocean coastline. It houses eight of the ten largest cities north of the Arctic Circle. And it is unapologetically assertive about its Arctic nature and territory and its right to encourage others to take advantage of the opportunities its Arctic waters present while being simultaneously selective about who it permits to do so.

In a 1991 essay in *The Soviet Maritime Arctic*, Franklyn Griffiths of the University of Toronto attempted to get to grips with the role of the Arctic in the Russian identity and found it to be as full as contradictions as, to many outsiders, Russia itself appears to be. Russians, he argued, "must be regarded as a northern people. They have powerful and historical attachments to a northern landscape that has been a primordial force in shaping their customs and awareness. . . . Russians, it follows, may be predisposed to northern activities in which they are able to express themselves as a people. They are less constrained than others by perceptions of climatic severity, higher operating costs and greater uncertainty. . . . By extension, therefore, it could be that Russians are more favorably disposed towards undertakings in arctic regions." In particular, he hypothesized, "Arctic marine disasters, bold rescues and demonstrations of ability to prevail in the midst of adversity all serve to endow Northern Sea Route activities with undoubted public appeal. Russians may not wish to live in the Arctic, but they're likely to derive vicarious satisfaction and a sense of common achievement from the spectacle of arctic marine operations and other precedent-setting attainments of . . . northern development."

And yet, as Griffiths himself noted, the situation is more complex and nuanced than Russians seemingly being culturally attuned and climatologically disposed toward Arctic matters. After all, prior to the twentieth century, Russian habitation of and involvement in its Arctic reaches was sporadic and fitful, for entirely understandable reasons; furthermore, he notes that when the Soviet Union did begin its expansion into Arctic realms, it did so in a way that sought not to accentuate the region's differences but to eradicate them, to "denorthify" the north and subsume it under a wave of industrial megaprojects, projecting southern ways and means onto the Russian Arctic. It was less a case of a naturally Arctic people returning home than "outsiders bent on gain" eager to "suppress natural resistance," their underlying predispositions "those of the conqueror or treasure-hunter contemplating his future prospects: material gain, power, vindication."

Indeed, Joseph Stalin's attitude toward the Arctic was not outwardly significantly different in its generalities from the kind of words one might hear from any eager politician in any Arctic state today. "The Arctic and our northern regions contain colossal wealth," he declared. "We must create a Soviet organization which can, in the shortest period possible, include this wealth in the general resources of our socialist economic structure."

Elizabeth Buchanan, an expert on Russia and the Arctic with the Australian National University's National Security College, has observed that Stalin's influence was key to the development of Soviet interests in the Arctic. Indeed, as she states in her book *Red Arctic: Russian Strategy under Putin*, Stalin saw clearly the strategic potential of the region and was personally involved in key signature projects in the region, including the growth of Arctic aviation and establishment of the Northern Sea Route. (As she also noted, he rather more infamously ensured that for many, the region became synonymous with his system of penal camps, or Gulag.)

The resource wealth of the Russian Arctic, she wrote, "promised to fund Stalin's industrialization dreams, and the harshness of the Arctic environment coupled with humanity's battle to conquer it was a neat fit for communist propaganda. . . . What emerged was the concept of the 'Red Arctic'—a myth of popular Soviet culture epitomizing the leadership of the Soviet Union in all matters Arctic."

However, she noted, with Stalin's death, interest in the Arctic faded and the region's significance to Moscow's policymakers mostly centered on it being a "strategic and military space," given that the Arctic was the shortest distance between east and west: a place that provided both opportunity and vulnerability, a place for Russians to show off their abilities and achievements to the outside world and yet simultaneously one from which the outside world needed to be excluded. In this context, Griffiths argued that it is interesting that Soviet authorities in the 1930s went to some lengths to come up with an acceptable official designation of the waters of the Northern Sea Route and points farther north:

> On 27 June 1935, the All-Union Central Executive Committee, precursor of the Supreme Soviet, decreed, inter alia, that the term "Arctic" would no longer be used to refer to the waters in question, which henceforth were to go by the name "Northern Ice-Covered Ocean" (*Sernyi Ledovityi okean*).
>
> Faced with a choice, Soviet authorities opted for a native Russian and not a Western expression, and for "North" as distinct from "Arctic." In declining to endorse generally accepted international usage, they would seem to have expressed an underlying determination not only to deny access by Westerners to Soviet land and offshore areas, but to deny the thought in Soviet Russia of shared access to a high-seas area denoted by a single generally accepted name. The Arctic Ocean was instead to be viewed as a high-seas area over most of which Soviet Russia exercised exclusive political control and over the remainder of which Soviet rights of access under international law would be unconstrained.

The arrival of Mikhail Gorbachev led to something of a sea change in that attitude, with the final Soviet leader proposing in 1987 a reduction in military presence in the region on both sides and offering the prospect of international access to the Northern Sea Route. That in turn prompted something of an Arctic détente, much as glasnost and perestroika led to a broader global one. Russia and Canada engaged in their first multilateral negotiations since the formation of the Soviet Union. Gorbachev's ministers visited Norway and Sweden and proposed reductions in the number of armaments in the Arctic region. Such cooperation extended through the breakup of the Soviet Union and the presidency of Boris Yeltsin, even through the chaos that economic "shock therapy" visited on Russia. The rise of Vladimir Putin, however, brought about a new age in relations between Russia and the West, one in which comity and comradeship steadily yielded to suspicion and in turn outright hostility, with the Arctic ultimately becoming caught in the crossfire.

———————

In 1926, the Soviet Union filed a territorial claim to much of the Arctic Ocean, a claim that encompassed the full length of its coast northward to the North Pole. No other country recognized it, and the claim gained no traction. Seventy-five years later, Russia submitted another one, this time based on an assertion that the country's continental shelf extends to the North Pole by way of an undersea mountain range known as the Lomonosov Ridge. The claim, which was modified in 2012 and 2015, prompted some anxiety among other Arctic nations and set off a series of similar filings by Canada, Denmark, Norway, and the United States as well as what would become a wave of stories about the prospect of a new Arctic cold war.

The narrative—burnished by comments such as Russian prime minister Dmitry Medvedev's declaration in 2008, "I want to especially underline that this is our duty, this is simply a direct debt to those who have gone before us. We must firmly, and for the long-term future, secure the national interests of Russia in the Arctic"—was to some extent shaped by context. Vladimir Putin acceded to the presidency in 2000, one year before Russia filed its claim, and notwithstanding then-president George W. Bush's 2001 assertion that he had looked Putin in the eyes, "found him to be very straightforward and trustworthy," and "was able to get a sense of his soul," the relationship between the Russian president and Western leaders has become progressively tenser, more distant, and more marked by mutual suspicion. To outside observers, Putin appears uncomfortably comfortable with, even keen on, Russia positioning itself almost reflexively in opposition to the West and pursuing policies that at least echo those from the Soviet era.

To Elizabeth Buchanan, that is not a misreading of the situation: "For Putin, the Arctic offers a throwback to Soviet times in which he grew up and about which he is strongly nostalgic," she wrote in *Red Arctic.* "The Arctic provides a new avenue for reinstating Soviet symbols in contemporary Russia." Putin's Arctic strategy, she continued, serves two goals: "to outline Russia's national interests in the Russian Arctic zone, and to articulate the threats and challenges posed in the region to Russian national security." And yet, she argued, this does not mean that Russia has any intention of behaving unilaterally or breaking away from any Arctic consensus; in fact, Russian Arctic strategy is "pragmatic, in line with international law in the region, and not dissimilar to the other Arctic rim powers." Moscow's relationship with rival nations may be conflict oriented in other realms, she wrote, but in the Arctic it has continued to act cooperatively—at least until its actions in those other spheres sharply proscribed the opportunity for cooperation.

In his book *Russia and the Arctic,* Geir Hønneland notes Russian arguments that any perceived Arctic bellicosity on Moscow's part in recent years is, as much as anything, merely a response to the tub-thumping of the early years of the Stephen Harper administration in Canada. He cites Russian media reports from 2010 that quote former Canadian foreign minister Lawrence Cannon as saying that "the Arctic has always been a part of us, it still is, and always will be"—an unremarkable comment from an Arctic nation looking to reinvigorate its commitment to the region but, within the context of Harper's early expressed intent to increase military spending in the region, and to suspicious ears trained to assume the worst from the West, a veiled threat. Hønneland further notes that Russia is acutely sensitive to the fact that of the five nations with Arctic Ocean coastlines, it is the only one that is not a member of the North Atlantic Treaty Organization; hence the expression in Russian media of concerns that the "USA and its partners in NATO are striving to extend their economic presence in the northern waters, and to achieve internationalization of the Northern Sea Route and, as a result, press Russia out of the region"—the latter a continuation of the insecurity that prompted Tsar Mikhail to close the Mangazeya Sea Route in the 1600s.

But if Russian Arctic strategy is suffused with a combination of paranoia and conflicting desires to both economize and protect the Northern Sea Route, what in specific terms are its goals? Buchanan notes that the 2008 Principles of the State Policy of the Russian Federation in the Arctic to 2020 and Beyond articulated four principal national security goals, including the utilization of the country's Arctic region as a "national strategic resource base capable of fulfilling the socio-economic tasks associated with national growth" and "the use of the Northern Sea Route as a unified transportation link connecting Russia to the Arctic." Subsequent plans have refined some points, underlined others, and de-emphasized yet others—but two fundamental elements remain.

One is addressing and overcoming "threats" to Russia's control of its Arctic region, meaning not just those from outside actors but also internal issues such as lack of infrastructure and development, and a small yet declining population, in the area. The other, both distinct from and fundamental to the first, is consolidation and expansion of the Northern Sea Route.

In pursuit of that second goal, Putin has encouraged modernization and development in the form of expanded and improved ports, communication and navigation infrastructure, and search and rescue capabilities, as well as through interest-free and tax-free incentives for energy, petrochemical, and related projects that can contribute to the NSR's payload. Such efforts have, on one level, proven productive: in 2023, more than 34 million tons of cargo passed through the NSR, a significant improvement on the 7 million tons that transited the passage in 1987. But that figure is not only a mere fraction of the 135.5 million tons that transited the Suez Canal in November 2023 alone; it is also some distance shy of the 80 million tons that Putin himself set as a goal for 2024. While some degree of shortfall was expected—Buchanan wrote that Putin's figure was "widely considered an unrealistic objective"—the fact of the matter is that when it comes to expanding ship traffic on the Northern Sea Route, Russia may face no greater obstacle than Russia itself.

CHAPTER 10

Regulations and Reservations

In the far northeast of Norway, on a peninsula of land along a fjord called the Bøkfjorden, which is itself the southern arm of a longer and wider fjord known as the Varangerfjorden, sits the small town of Kirkenes.

Kirkenes has a population of just under 3,500—which swells to nearly 8,000 if including the surrounding suburban villages of Hesseng, Skytter-husfjellet, Sandnes, and Bjørnevatn. It is so far east that Finland, nominally Norway's eastern neighbor, is to its southwest; traveling from Kirkenes to Helsinki actually involves moving *forward* an hour instead of backward, as would normally be the case with a westward journey.

Kirkenes is 250 miles north of the Arctic Circle, farther east than Istanbul or Saint Petersburg, approximately 1,100 miles from Oslo but only 130 or so from Murmansk. The closest Norwegian town to Russia, it lies just a couple of miles away from the border; in 2010, the two governments signed an agreement that allowed those living in the Kirkenes area to cross into Russia and spend up to fifteen days at a time within thirty kilometers (nineteen miles) of the border without a visa, while Russians from the border region could spend the same number of days within the same distance on the Norwegian side.

Russian is spoken as widely and frequently in Kirkenes as is English, and many street and store signs are in Russian as well as Norwegian. Russians have traditionally visited Kirkenes to shop for goods not easily available on their side of the border, while Norwegians have driven into Russia to top up on cheap fuel.

As with Murmansk, the harbor at Kirkenes is ice-free year-round. Its waters are up to thirty meters (ninety-eight feet) deep; it hosts fishing ves-sels that call before and after excursions into the Barents Sea; and it boasts extensive dry dock facilities—all of which would appear to make it the ideal international gateway to the Northern Sea Route.

In 2019, a Finnish company, FinEst Bay Area Development, said it planned to build a railway between Kirkenes and northern Finland to take advantage of the seemingly inevitable increase in Northern Sea Route traffic and concomitant growth in the Norwegian port. It estimated that the undertaking would cost a total of $3.4–$5.6 billion, although the previous year a Finnish government study concluded that such a project would take fifteen years to complete and would not be commercially viable. Also in 2019, a festival in town sought to highlight the prospect of Chinese investment in the area and of becoming the world's "northernmost Chinatown." That followed a sudden surge in Chinese tourism to the area and growing expressions of interest in investment in the town. In 2010, the first non-Russian ship to transport non-Russian cargo through the NSR was the *Nordic Barents*, which carried iron ore from Kirkenes to Lianyungang in China.

Even with uncertainty over the immediate viability of the Northern Sea Route, Kirkenes appeared unmatched in its ability to take advantage of the increased trade that appeared inevitable. It seemed merely a matter of when, not if, Kirkenes would be able to blossom into a major international port.

And then, on February 24, 2022, Russia invaded Ukraine, and everything changed.

In 1596, a Dutch expedition led by Willem Barents and Jacob van Heemskerck sailed north from Norway, became the first Europeans to spy Svalbard, and pushed east in an attempt to locate and navigate the Northeast Passage. They sought to round the archipelago of Novaya Zemlya but, thwarted by ice in the Kara Sea, elected to spend the winter in a shallow bay they called Ice Haven. The name proved overly optimistic; although they had hoped to stay onboard their vessel, ice squeezed the hull even in their putative sanctuary, forcing them to abandon ship and build a shelter ashore. All but one of the crew survived the winter, but they had to contend with the attentions of a succession of polar bears, including a pair that stalked men who were working outside and one that, in the depths of winter, pushed forcefully on the shelter's door. Nor, for many of the men, was it their first experience with polar bears in the area; on another voyage two years before, a bear had surprised some of the men and killed two of them, prompting most of the rest, not unreasonably, to run off in fright.

More than four hundred years later, in February 2019, an "invasion" of more than fifty polar bears left the inhabitants of Novaya Zemlya "scared to send their children to school." The archipelago's principal settlement, Belushya Guba, was placed under a state of emergency after the bears, driven ashore by a lack of sea ice, entered the community to scavenge at the waste dump and around people's homes and even "literally chased people in the region."

In the several centuries between those two events, the archipelago underwent a wave of Russian colonization in the late 1800s; it was an important base in World War II, from which the USSR's Northern Fleet sought to prevent Nazi warships from entering the Northern Sea Route (and which was the target of a U-boat attack in August 1942); and, in the 1950s, it was largely denuded of human habitation as it was turned into a major military station. In particular, Novaya Zemlya was one of the Soviet Union's two principal sites for nuclear weapons testing. It was here in October 1961 that the Soviets detonated Tsar Bomba—or Emperor of Bombs—which at approximately fifty megatons was an astonishing 2,380 times more powerful than the bomb that was dropped on Hiroshima and was, in fact, the largest bomb ever detonated.

Today, Belushya Guba, the archipelago's principal settlement, has around 2,500 residents, most of them military or construction workers, and access to Novaya Zemlya remains restricted, with visitors needing special permits. Prior to the Ukraine invasion, however, some passenger ships regularly secured permission to visit as part of cruises through the Northeast Passage. Passengers might step ashore at Glazov Bay on the west coast of Severny Island, the northernmost of the two major islands in the archipelago, to drink in the views of mountains and glaciers and look for bird life such as eiders and terns, or to stop by Ice Haven, where only a few pieces of scrap timber survive of the shelter in which Barents, van Heemskerck, and crew sought refuge from the Arctic winter and ravenous polar bears. Alternatively, they might head south and land on Ostrov Vaygach, home now to approximately five hundred Indigenous Nenets people and more than four thousand reindeer.

Stretching more than 600 miles from northeast to southwest, and measuring between 40 and 90 miles in width, Novaya Zemlya is a striking and unmistakable feature of the Russian Arctic coastline. Approximately one-quarter covered by glaciers, with a mountain range that runs the majority of the length of its spine, it sits northeast of Murmansk and yet is in many ways the true western entry point into the Northern Sea Route. It is also the first route marker you see when traveling from west to east: turn north of Novaya Zemlya and you take one path through the NSR; duck south and you take another.

In broad terms, the northern sea lane follows a route that runs north of all the major islands and archipelagos in the passage: north of Novaya

Zemlya, north of Severnaya Zemlya, north of the New Siberian Islands, and then via the Long Strait, which separates Wrangel Island from the Siberian mainland, into the Chukchi Sea and the Bering Strait. The more southerly route runs via the Kara Strait south of Novaya Zemlya, the Vilkitsky Strait south of Severnaya Zemlya, the Sannikov Strait south of the New Siberian Islands, and the Long Strait. The northern sea lane is 230 miles shorter than the southern sea lane, but ice tends to remain in the northern sea lane even in summer, while currents tend to pull ice away from shore along the southern route, theoretically leaving a navigable pathway. Depending on the ice conditions, ships may take a route that uses an appropriate combination of the northern and southern sea lanes.

While sea lanes in the Northern Sea Route tend to be broader and deeper than the narrow, twisting, and frequently shallow pathways of the Northwest Passage, they are not, to coin a phrase, all plain sailing. Taking the southern path requires navigating a series of straits that can be as deep as 650 feet but as shallow as 33 feet and presents several potential ice-filled choke points: at Novaya Zemlya; at the archipelago of Severnaya Zemlya, approximately halfway along the NSR's length; at the New Siberian Islands; and around Wrangel Island. Although the navigational season in the NSR traditionally runs from early July to mid-November, with the prospect of it extending more frequently through December as climate change affects fall and winter ice formation, the beginning and end of that season can still be particularly treacherous. In early November 2021, eighteen vessels became stuck in the ice for several weeks at five points along the passage following a much earlier and more rapid freeze than anticipated; the intervention of several nuclear-powered icebreakers was required to free them. And while the summer months at least theoretically offer a greater number of possible ice-free days than winter, they are simultaneously prone to thick fog, particularly near the edge of concentrated ice, where cold air meets warmer water or warmer air clashes with ice.

Partly in response to such safety concerns, Russia strictly regulates the type of shipping that can enter the NSR. Commercial and adventure vessels must first submit an application, no earlier than four months and no later than fifteen days before the planned date of arrival. If permission is granted, entry must not be any earlier than the assigned date, and the vessel's master must notify Russian authorities when it is 72 hours and 24 hours away.

For navigational purposes, Russia divides the Northern Sea Route into twenty-eight areas, and whether a ship is allowed to operate with or without an icebreaker, or whether it requires an ice pilot, depends not just on the extent of ice strengthening of the ship but also on ice conditions at the time in the areas in which it will be traveling. Ships with minimal or no ice strengthening are prohibited entirely from the NSR from mid-November to June; oil tankers, gas carriers, and chemical carriers with a gross tonnage in excess of ten thousand tons may travel in areas of open water only from July to November 15, and only with icebreaker assistance; other ships without ice strengthening can operate without icebreaker assistance only in areas where the water is open from July to mid-November.

It is not always the case that a ship gains an icebreaker all to itself. More common is that ships are lined up in a convoy with one or more icebreakers leading the line and breaking channels through ice as appropriate while maintaining regular communication with the vessels behind it.

If that all sounds straightforward and efficient in principle, in practice it has not always been thus, and would-be transiters have at times found themselves frustrated by authorities' apparent capriciousness and by the seemingly ongoing internal Russian tension between welcoming international trade and instinctively erecting as many barriers as possible to foreigners.

David "Duke" Snider, founder of Martech Polar Consulting, which provides ice pilotage and navigation services for vessels operating in the Arctic and Antarctic, understandably has not conducted any business in the Northern Sea Route since the invasion of Ukraine; but, he said, in the years prior he found that "the bureaucracy that you have to go through to be approved for a voyage, let alone anything to pick up cargo, is a massive nightmare. Besides the slow process and everything else like that, there is the penalty you pay right off the bat when they say 'OK, well, you're gonna have to pay for icebreaker support.' They don't say that in their press releases about creating a highway across the north. The last time we were involved in the Northern Sea Route, it was going to be about $400,000 just to secure permission to ask for an icebreaker."

According to a 2020 analysis by Jan Jakub Solski published in the *Arctic Review on Law and Politics*, Russia has at least attempted to approach the issue of permitting and fees with increasing consistency and transparency as use of the NSR has become more established and frequent. In 2013, there were 718 applications to enter the Northern Sea Route, of which 83 were denied at first time of asking. In 2014, 30 of 661 applicants were turned down;

in 2015, 15 of 701; in 2016, just 3 of 721. Sixty-five of those 83 rejectees in 2013 were granted permission after fixing their applications and submitting them a second time; the following year, 24 of the 30 who were initially turned down were granted a reprieve at the second time of asking. In 2015, 6 of the ships that were rejected on first submission ultimately received permission to enter the NSR; of the other 9, 3 were rejected because of being insufficiently ice strengthened at a time of "dangerous [ice] conditions." One—a Russian ship—had already completed its voyage without a permit, while the others either did not reapply, did not provide sufficient information, or wanted to operate outside the areas that their ice classification permitted. Of the 3 that were turned down the following year, 2 were successful with a follow-up submission after fixing some basic clerical errors in their forms.

One ship that was not at any stage granted a permit was the Greenpeace icebreaker *Arctic Sunrise*—the same vessel, incidentally, on which I sailed up the Bering Strait to Little Diomede Island and toward Wrangel Island in 1998—even though it clearly satisfied the icebreaking criteria. If ever there were a clear example of a ship being denied entry on political grounds, this would be it; four times the *Sunrise* applied and four times it was rejected. In September 2013, following the third rejection, the ship sailed into the Pechora Sea, to the south of Novaya Zemlya, and activists onboard attempted to scale the Prirazlomnaya drilling platform as part of the organization's ongoing protest against oil drilling in the Arctic. Two succeeded, only to be removed and detained onboard a coast guard vessel; the following day, Russian authorities seized the *Arctic Sunrise* and towed it to Murmansk. All thirty onboard, including two journalists, were beaten and interrogated and held in detention until December, when they were released, purportedly as part of a general amnesty to mark the twentieth anniversary of the Russian constitution. The ship itself was not released until the following June.

The *Arctic Sunrise* was sailing under the Dutch flag, and the Netherlands formally objected to its seizure, noting that it was outside both Russian territorial waters and a safety zone around the rig and that Russia was bound by international law to contact Dutch authorities before conducting any operations against the vessel. In August 2015, the international Permanent Court of Arbitration ruled that Russia had violated the Law of the Sea and that Russian claims of piracy and hooliganism did not apply. (Russia responded by stating that it did not recognize the tribunal's authority. It had justified its actions on the grounds of "violation of the Rules of navigation in the water area of the NSR, adopted and enforced by the Russian Federation in accordance with the article 234 of the United Nations Convention on the Law of the Sea.")

The article 234 that Russia cited applies specifically to navigation in polar regions and forms the basis of much of Moscow's national legislation and guidance on the Northern Sea Route. It was included in discussions of the United Nations Convention on the Law of the Sea (UNCLOS) at the behest of Canada, which sought international legal support for its Arctic Waters Pollution Prevention Act (AWPPA), passed in 1970 as part of its efforts to control international traffic in the Northwest Passage. Specifically addressing navigation in ice-covered waters, the AWPPA provides that "coastal States have the right to adopt and enforce non-discriminatory laws and regulations for the prevention, reduction and control of marine pollution from vessels in ice-covered areas within the limits of the exclusive economic zone." Its language is, as Solski's paper notes, "ambiguous," as a result of which it was able to "satisfy multiple delegations with divergent interests," and it formed the basis of a revision to Russia's ever-evolving legal standards for the Northern Sea Route.

The first such standards were codified in the 1990 Regulations for Navigation on the Seaways of the Northern Sea Route, which were supplemented in 1996 by regulations on icebreaker and pilot guiding and on the design, equipment, and supplies of vessels navigating the NSR. Broader rules that incorporated the NSR followed, including the 1998 Federal Law on the Internal Maritime Waters, Territorial Sea and Contiguous Zone of the Russian Federation (conveniently abbreviated to IWTSCZ) and the 1999 Merchant Shipping Code of the Russian Federation. The latter served as the legal basis for establishment of the Administration of the Northern Sea Route (ANSR), which is the body that reviews and grants—or denies—applications for permits to sail through the NSR. The 1999 code includes parameters for the regulation of navigation, operation of the ANSR, permits, and fees and formed the basis of the most recent major overhaul of regulations, the 2013 Rules of Navigation in the Water Area of the Northern Sea Route. Those rules established firm, fixed boundaries for the NSR, including its delimitation at the edge of Russia's two-hundred-mile exclusive economic zone; in addition, by dint of their being anchored in the 1999 Merchant Shipping Code, they apply solely to commercial ships, whereas the 1990 regulations offered no distinction between such ships and state-owned vessels. This was always likely to prove resistant to any legal challenges, as the Law of the Sea states explicitly that the provisions of the law regarding the protection and preservation of the marine environment—such as the aforementioned article 234—"do not apply to any warship, naval auxiliary, other vessels or aircraft owned or operated by a State and used, for the time being, only on government non-commercial activities."

This effectively left the status of warships, and the ability of foreign navies to navigate the passage, in something of an abeyance until December 2022, when Vladmir Putin signed an amendment to the 1998 federal law on IWTSCZ specifically to address it. The new regulations require a flag state to apply for permission to enter NSR internal waters ninety days before passage, with no more than one warship to be allowed entry at any one time unless the Russian government provides special dispensation. Submarines are required to surface and display their flag while traveling through the passage.

In an analysis of the rules for the Harvard Kennedy School's Belfer Center for Science and International Affairs, Andrey Todorov noted that "some Russian media praised the measure as a showcase of Russia's growing dominance in the Arctic and its ability to 'wipe the nose' of its adversaries, while Western commentators pointed to it as more evidence of Russian arbitrariness and disregard for international law." But he emphasized that the rule applies specifically to internal waters—those within twelve nautical miles (approximately fourteen statute miles) of the coast—which has long been accepted as a legitimate requirement for coastal states. (He added that an earlier draft of the legislation would have required flag states only to provide notification and not solicit permission but would have applied to a much broader area; he speculated that Moscow likely rejected this specifically because it would have been extremely hard to defend under international law.) The bone of contention, however, reflects the dispute over the Northwest Passage, in that Russia includes in its internal waters the various straits through which a ship must pass if taking the southern route through the NSR—which, as we saw earlier, because of ice conditions is frequently the only navigable path. For that reason, the United States maintains the same objection that it does with Canada: that those straits should be considered international maritime rights-of-way.

Todorov wrote that "some commentators contend that the adoption of the Act gives the United States cause for conducting a Freedom of Navigation Operation (FONOP) in the NSR." FONOPs, he pointed out, are naval operations that the United States conducts to "demonstrate its non-acquiescence to maritime claims by other states that it considers excessive." For example, on November 3, 2023, the USS *Dewey* sailed through the Spratly Islands in the South China Sea, which are contested by the People's Republic of China (PRC), Taiwan, Vietnam, Malaysia, and the Philippines (and, in part, Brunei), several of which impose unilateral restrictions on vessel traffic despite their location in the midst of strategic shipping lanes.

However, while the United States has conducted FONOPs *near* the Arctic, it has not done so *in* the Arctic, and it is questionable whether it could do so even if it wanted to, given that it does not have any ice-strengthened warships. (Imagine, wrote Todorov, if the United States attempted such an operation and then had to call on Russian icebreakers for assistance.) Further, given increased tensions between the two nations as a result of the Ukraine invasion, and the clear importance that Russia assigns to the NSR, it is questionable whether the dispute is worth the level of escalation that such an operation might provoke. Better for now, perhaps, to continue agreeing to disagree and to allow the dispute to simmer quietly in the background, as it has done for many years.

———⌇———

From the time of the establishment of the Glavsevmorput in the early 1930s, the Northern Sea Route has been a regular shipping corridor for goods, supplies, fuel, and equipment along the Arctic coastline of Russia and, before it, the USSR. It expanded into a transport route for raw materials such as timber, coal, and minerals from production sites near the coast, with a mining and metallurgical facility in Norilsk, near the Kara Sea coast, becoming the single most important cargo producer for the NSR. Indeed, the Soviet nuclear icebreaker fleet was developed largely to escort traffic to and from Dudinka, the loading port for Norilsk.

Not until January 1, 1991, was the passage formally opened to international trade, and the start was a fitful one. Between 1993 and 1999, Russia conducted a research project—the International Northern Sea Route Program (INSROP)—to explore the NSR's feasibility as a major trade route, as a part of which a vessel from Norilsk Nickel's cargo fleet, *Kandalaksha*, sailed from Yokohama, Japan, to Kirkenes in August 1995. The initial conclusion of INSROP was not encouraging: the prevalence of severe sea ice conditions was simply too great for regular international commercial shipping to be viable. While the door was theoretically open for other countries to ship goods through the NSR, the invitation was rarely accepted.

But growing realization of the extent of Arctic sea ice loss spurred greater interest. In a separate study published in the *Arctic Review on Law and*

Politics, Norway's Björn Gunnarsson of Nord University and Arild Moe of Fridtjof Nansen Institute noted that in 2005, the Arctic Council published its Arctic Climate Impact Assessment, which highlighted sea ice loss and other changes as a result of global warming; the summer and fall of 2007 saw what was then by some distance the lowest Arctic sea ice minimum on record; and in 2009 the Arctic Council's Arctic Marine Shipping Assessment focused attention on the prospect of long-sought passages through the ice.

The first true decade of international shipping on the NSR, Gunnarsson and Moe therefore argued, was the 2010s, and during that decade there were 89 international transit voyages through the passage—that is, journeys that began and ended outside the Northern Sea Route and did not stop along the way. There was tremendous variance from year to year; the lowest number of such transits in any twelve-month period was 1 in 2010—the journey of the *Nordic Barents* from Kirkenes—while the highest was 17 in 2018. The trend over the course of the decade, however, seemed clear: while there were 46 transits over the first seven years, there were 43 combined in 2017, 2018, and 2019 alone.

A major factor in that late-decade growth was China's COSCO Shipping Corporation, the world's third-largest shipping company, which first sent a ship along the route in 2013 and increased its presence thereafter, with 2 transits in 2015, 6 in 2016, 5 in 2017, and 8 in 2018. In 2021, of 26 voyages to China via the NSR, 14 were operated by COSCO. A report by Malte Humpert on the news site *High North News* underlined COSCO's apparent commitment to exploring the potential of the NSR, citing an internal company presentation highlighting that the company's voyages through the passage had "reduced sailing time by 280 days resulting in a reduction of roughly 27,500 tons of CO_2—equivalent to the annual per capita CO_2 emissions of 3,300 Norwegians."

Those kinds of figures weren't exactly likely to keep the operators of the Suez and Panama Canals up at night—candidly, they would barely be regarded as rounding errors in the more established shipping routes. But the trend, in a short period of time, was clearly upward; Humpert quoted Michael Byers, Canada Research Chair in Global Politics and International Law at the University of British Columbia, as emphasizing that "the COSCO voyages indicate a serious intent on the part of the Chinese government, via a state-owned company, to take advantage of increasing Arctic shipping routes."

And then, in 2022, the first season after Russia's invasion of Ukraine, international transits through the NSR cratered. That year, not a single ship conducted an international transit as economic sanctions and the fear of being punished for violating them took hold. "The feeling among international shippers and traders is that everything that goes through Russia now is like acid," Russian Arctic expert Mikhail Grigoriev was quoted as saying by *High North News*.

And yet, just one year later, the tune appeared to have changed. "Russia's Northern Sea Route Sees More Traffic Despite War and Sanctions," *High North News* reported on January 18, 2023; "China Pushes Northern Sea Route Transit Cargo to New Record," it declared on December 18 that same year. "Russia Claims New Record for Cargo on the Northern Sea Route," added the *Maritime Executive*, elucidating that Russian authorities reported that "they surpassed 35 million tons of cargo transported on the route in 2023."

So, what happened? Did fear of sanctions or disapproval of Russia's foreign adventures disappear overnight? The devil, as so often, is in the details—and specifically in the nature of the voyages that took place and the cargo that was transported.

Discussion of the Northern Sea Route, and the Northeast Passage more broadly, becoming a major trade route have focused—in this book and elsewhere—almost exclusively on international transits, on shipping companies seeing it as the pathway between Pacific and Atlantic that explorers had sought for centuries. But even as such transits grew during the 2010s, they remained a minority, in terms of both total voyages and cargo carried. As Gunnarsson and Moe noted, the route sees multiple kinds of voyages, including the following:

- *International transit voyage on the NSR*: shipping between two non-Russian ports that passes through both the western and eastern boundaries of the NSR.
- *Destination voyage on the NSR*: shipping between a Russian port and a non-Russian port. (A destination voyage can also be a transit voyage, e.g., from Murmansk to Yokohama.)
- *Domestic voyage on the NSR*: shipping between two Russian ports.

While international transits have collapsed, domestic and destination voyages remain strong—indeed, in many respects stronger than before as Russia adapts to changing circumstances.

Domestic traffic is now dominated by the Yamal LNG project, a massive liquefied natural gas plant in northwestern Siberia; groundbreaking for the plant began in 2012 and construction started in 2013, with the transport of machinery, equipment, and infrastructure until the plant came online in 2017 boosting total cargo transport figures on the NSR. Progress on a second plant, the Arctic 2 LNG project, delayed by a year as foreign investors withdrew because of Ukraine-related sanctions, resumed in 2023 before being scaled back the following year. Yet another development, the Vostok Oil project, just north of Norilsk, is projected to produce one hundred million tons of oil and requires "construction of unprecedented amounts of infrastructure objects."

Meanwhile, destination shipping continues apace; indeed, when the European Union banned the import of Russian crude oil at the end of 2022, the country emphasized the use of the Northern Sea Route to ship crude to markets in Asia, particularly China. Shipments of iron ore, coal, liquefied natural gas, and generalized cargo combined with shipments of crude to nudge overall cargo shipments to new heights. Meanwhile, Russia has announced its intention to spend $30 billion on continuing to develop the Northern Sea Route by 2035.

Outwardly at least, then, Moscow remains bullish on the NSR's role and potential. Indeed, if anything, as Russia has become more isolated as a result of its actions, and as other trade corridors and trading partners become closed off, its reliance on its best geopolitical friends consuming its goods and materials increases, and so does the importance of the Northern Sea Route as the means of delivering them. With hundreds of vessels of various sizes servicing several massive infrastructure projects in the Arctic and exporting millions of tons of cargo, it is undeniably an impressive enterprise, a culmination of the Stalinist goal of "denorthification."

But the ultimate goal—of the Northern Sea Route being that long-dreamed-of trade route between East and West—feels more distant than ever. And meanwhile, curious and covetous eyes find themselves drifting even farther north.

North

CHAPTER 11

No Sea Unnavigable

Even as waves of explorers probed the waters of northern Europe and North America in search of passages to the northeast and northwest, others set their sights still farther north, to the very top of the world and beyond, picturing a pathway that would traverse the very top of the world.

Robert Thorne, an English merchant living in Seville, found the notion so compelling that he sent a letter to that effect to King Henry VIII in 1531. Thorne argued that a route from Europe to Asia that went directly north would be almost 2,000 leagues (approximately 7,000 miles) shorter than those around Cape Horn and the Cape of Good Hope. He provided a map that purported to show, notwithstanding the lack of supporting evidence, that the North Pole was surrounded by open sea that would allow for safe transit, and he argued that the most treacherous parts of any such voyage would be the 300 leagues immediately prior to and after the attainment of the Pole.

Thorne's entreaties were not acted upon by the king, but the notion persisted for centuries, alternately losing and regaining favor as the latest discoveries and speculations either supported or negated it. In 1607, the Muscovy Company dispatched Henry Hudson to find a route across the Pole onboard the *Hopewell*; he reached Spitsbergen and crossed north of 80 degrees before being forced to turn south. In 1773, a British Admiralty expedition under the leadership of Constantine John Phipps achieved much the same results; in 1806, William Scoresby, also seeking a path via Spitsbergen, attained 81°30', the most northerly latitude officially reached by a vessel to that point, but could reach no farther. Before leading the expedition that would result in his death and that of his crew, Sir John Franklin supported David Buchan on another Admiralty venture that took a similar route and reached a similar latitude before encountering similarly impenetrable ice. Both of the expedition's ships experienced severe damage as a result of storms and ice; the Admiralty would not attempt to sail north across the Pole again.

The idea gained traction on the other side of the Atlantic, however, where in the mid-1800s its disciples promoted it with an almost messianic fervor. Elisha Kent Kane, who had been one of the discoverers of the Franklin expedition graves on Beechey Island, left New York in May 1853 onboard the *Advance*, in command of an expedition with the twin aims of making another attempt to find Franklin and sailing north of Baffin Bay in search of the Open Polar Sea. He succeeded in neither goal; but, sailing up the west coast of Greenland, he pushed as far as what is now known as Kane Basin, an at best episodically open bay in Greenland's extreme northwest, before becoming beset for the winter. The ice would never loosen its grip, holding the *Advance* tightly during that winter, the succeeding summer, and the following winter, until in May 1855, the expedition members—who had been kept alive largely through the benevolence of nearby Inuit—set out on foot, again with Inuit assistance, for the Greenlandic community of Upernavik. They reached their goal in August, and by September, they were on their way home, courtesy of a relief expedition.

Despite the immense privations suffered during the endeavor, which saw three of the twenty company and crew perish, the expedition's members returned with, if anything, an even stronger belief in the existence of an Open Polar Sea, not least because some of them claimed to have seen it. During the summer of 1854, William Morton and Hans Hendrik sledged north through Kane Basin and, according to Morton, spied in the near distance an expanse of water that contained "not a speck of ice. . . . As far as I could discern, the sea was open. . . . The wind was due N(orth)—enough to make white caps, and the surf broke in on the rocks in regular breakers."

Morton and Hendrik had come across a polynya, a short-lived area of open water amid the sea ice that is a frequent feature of polar waters; but Morton's description was sufficiently vivid and convincing that upon his return he was dubbed the "discoverer of the Open Polar Sea."

Indeed, in the first ever comprehensive work on oceanography, *The Physical Geography of the Sea*—published shortly after Kane's return—author Matthew Fontaine Maury wrote breathlessly, and with a tenuous relationship to the facts:

> Seals were sporting and water-fowl feeding in this open sea of
> Dr. Kane's. Its waves came rolling in at his feet, and dashed with
> measured tread, like the majestic billows of old ocean, against
> the shore. Solitude, the cold and boundless expanse, and the

mysterious heavings of its green waters, lent their charm to the scene. They suggested fancied myths, and kindled in the ardent imagination of the daring mariners many longings.

Not everybody was sold on the idea, however, and among those who disputed that Kane's expedition could have encountered an Open Polar Sea was German geographer August Petermann—not, however, because he was a skeptic about the sea's existence; far from it. Rather, he argued that Kane's location was all wrong; the sea, he argued, had to be located farther north and to the west because its existence was fueled by a warmwater current that flowed north from the Pacific through the Bering Strait.

Petermann's thesis intrigued James Gordon Bennett Jr., publisher of the *New York Herald*, who had recently found renown when he dispatched his reporter, Henry Morton Stanley, to Africa to find British explorer David Livingstone. Bennett had already funded one of the final searches for Franklin when, in 1877, he resolved to sponsor an expedition to test Petermann's theory. He would charter a ship to sail north through the Bering Strait; if it did not immediately find a passage to the polar sea, it would be tasked to explore the little-known Wrangel's Land—which we know to be an island off Russia's Arctic coast but which Petermann theorized was part of a landmass that stretched all the way to Greenland and could provide an alternative land-based path to the North Pole. One way or the other, Petermann assured Bennett, his bases would be covered.

Bennett procured a former Royal Navy vessel, the *Pandora*, which after being retired from naval service had made two exploratory voyages to the Arctic; he renamed it the *Jeannette*, after his sister, and placed it under the command of an active US Navy officer, George W. De Long. Although financed by Bennett, the voyage was subject to naval laws and discipline, was officially dubbed the U.S. Arctic Expedition, and set out from San Francisco to much fanfare on July 8, 1879. After making stops in the Aleutian Islands and Siberia, it turned north for Wrangel's Land on August 31; five days later, approximately fifteen nautical miles from Wrangel Island's neighbor, Herald Island, the *Jeannette* became trapped in the ice.

It would never escape.

After a month of being carried back and forth helplessly in the drifting ice, De Long and his crew realized, to their profound disappointment, that there was no Wrangel's Land; at the same time, measurements showed that there was no warm current flowing this far into Arctic waters, either. Already, those onboard were coming to grips with the realization that the entire rationale on which their endeavor was based was, at best, flimsy; and yet, there was nothing they could do to break free of their tomb or improve their fate. The dour mood that descended on the *Jeannette* was only worsened by the always present sight of Herald Island, reminding them of their failure and seeming to mock them as the ice carried them back and forth for weeks and months on end. In March 1880, De Long recorded that the ship's position was essentially the same as it had been three months before; after sixteen months, the ship had moved little more than two hundred nautical miles.

Suddenly, on June 11, 1881, the ice released its grip, and *Jeannette* was afloat in a small pool of open water; the next day, however, the ice closed in with a vengeance, crushing the hull and forcing De Long to order the vessel's abandonment. On June 13, the ship sank, and the crew set out—first across ice and then in three boats through tempestuous waters—for safety. One of the boats, containing eight men, disappeared at sea; the other two made landfall but had become separated, and a further twelve men died over the course of several weeks of fruitless slogging through harsh Siberian conditions. Just thirteen of the original complement of thirty-three men survived and returned to the United States, their tragic tale finally closing the book on the notion of a passage through the ice across the North Pole.

Or so it seemed.

————✢————

One hundred and thirty years after the *Jeannette* was crushed in the ice, seemingly taking dreams of a passage across the northern polar ice cap with it, a 2019 entry in the blog *Cryopolitics* addressed what its headline referred to as "the Arctic shipping route no one's talking about." Talk of the potential of the Northern Sea Route, wrote Mia Bennett of the University of Washington, was "starting to seem passé in the face of rapid climate change in the Arctic. The shrinking and thinning of sea ice is happening faster than scientists thought possible—so fast that now, it's not just the Northern Sea Route or even the Northwest Passage that people are talking about. They're talking about a trans-Arctic passage cutting straight across the North Pole."

The following year, Bennett and colleagues expanded on her article with a paper in the journal *Marine Policy*. A Transpolar Sea Route (TSR) could, they wrote, have several advantages over either the Northeast Passage or the Northwest Passage. For one, notwithstanding disputes and disagreements over who "owns" the North Pole—with several different countries, as we shall see, making territorial claims on at least some of the polar basin—any such route would, by dint of its distance from the coastlines of any of those nations, qualify as being in the high seas and thus exempt from national jurisdiction or, say, the icebreaker escort fees that Russia charges for shipping using the Northern Sea Route. (This is true more or less. Some entry and exit routes—particularly through the Bering Strait—would come within a matter of miles of one or more countries' coasts, while the Law of the Sea does allow nations to impose environmental regulations on ice-covered zones in their exclusive economic zones, which extend two hundred nautical miles from their coastlines. The interpretation of such regulations, let alone their implementation, in such a circumstance is far from clear, however, and would anyway apply, likely with greater stringency, to both the Northwest and Northeast Passages.)

The bathymetry (that is, the depth and morphology of the seafloor) would also be more attractive to the largest of ships than that of either alternative route. The entrance to the Laptev Strait, a key eastern portal into the Northern Sea Route, is just thirty feet or so deep, too shallow for the very largest container ships, such as the Very Large Crude Carrier (VLCC) and Ultra Large Crude Carrier (ULCC) vessels, which are also too mammoth to transit the Panama Canal. The TSR, in contrast, would offer no such limitations: its only real choke point, the Bering Strait, has a depth of up to 160 feet and can accommodate all but the very largest ULCCs. And the average depth of the polar basin itself is 3,400 feet.

Furthermore, the TSR would be at least competitive with the alternatives in terms of length. By one estimate, it measures approximately 3,760 nautical miles, compared with roughly 2,790 for the Northern Sea Route and over 5,000 for the Northwest Passage. The same analysis concluded that sailing from Tokyo to Rotterdam across the North Pole would entail sixteen days of sailing at a steady seventeen knots over a distance of 6,600 nautical miles, substantially shorter than the twenty-seven days and more than 11,000 nautical miles via the Suez Canal. (That advantage is, naturally, diminished or nullified the farther south one end of the journey is located. Singapore to Rotterdam, for example, is a shorter distance via the Northern Sea Route than across the top of the world.)

Such renewed enthusiasm for the prospect of a transpolar passage is an extremely recent phenomenon. As Bennett noted in her *Cryopolitics* blog entry, Norway's Arctic Strategy, updated in 2017 and 2020, doesn't mention it at all, although it does reference the Northeast Passage. Nor does Iceland's policy, adopted in 2011, although a 2021 update acknowledges it in passing. South Korea, which is increasingly playing an active role in Arctic matters, especially those regarding shipping, refers only to the Northeast Passage in its policy; Canada's, understandably, discusses only the Northwest Passage. A 2019 report to the United States Congress on changes in the Arctic noted only the Northwest Passage and Northern Sea Route; it also, Bennett observed, included a map of sea ice loss that was a decade old—representing, she wrote, "a time when there were only two conceivable Arctic Ocean routes rather than three."

That comment highlighted the prime reason why the notion of transpolar passage, so long dismissed as a fanciful conceit in the face of unrelenting ice, has so swiftly come once more to the fore. The ice that repelled Scoresby and Kane and crushed the *Jeannette*, the ice that is the defining feature of the Arctic Ocean basin, is disappearing far more rapidly than anyone had dared imagine.

The sinking of the *Jeannette* was enough to dissuade the last true believers of the existence of a readily accessible Open Polar Sea. But when, three years after the expedition's demise, relics from the ship's wreckage were discovered on the coast of Greenland, far from where the *Jeannette* had perished off the Siberian coast, it lit a spark in the minds of two Norwegians. The first, Henrik Mohn, founding director of the Norwegian Meteorological Institute, argued in an 1884 lecture at the Norwegian Academy of Sciences and Letters that the location of the relics showed that, whether or not there was an open sea, there was certainly an ocean current flowing from east to west across the Arctic Ocean. That caught the attention of Fridtjof Nansen, a curator at the Bergen Museum, who, although just twenty-three years old, was already preparing to lead the first crossing of Greenland, a feat he would accomplish five years later.

In February 1890, following the success of his Greenland mission, Nansen stood before the Norwegian Geographical Society and emphasized that the discovery of not only the *Jeannette* relics but also Siberian and Alaskan driftwood along the Greenland coast were strong indicators of the presence of a transpolar current. Furthermore, he planned to use it to carry him to the North Pole.

Whereas previous polar expeditions had sought to do battle with the ice, a conflict they could not possibly hope to win, Nansen would take a different tack. He would sail a ship as far north as he could until the ice entombed it, and then he would allow the current to carry the vessel, along with the ice pack, across the polar basin via the Pole until it emerged on the other side. His plan was met with almost equal amounts of enthusiasm and vituperation, the latter expressed with particular pithiness by American explorer Adolphus Greely, who criticized the notion as "an illogical scheme of self-destruction." Nansen, however, was unmoved and pressed ahead with plans to design and build a ship specifically suited for the task ahead. The design he settled on was for a vessel of unusually rounded appearance, the idea being that such a shape would ensure that rather than being squeezed between ice floes, it would "slip like an eel" out of the ice's deadly embrace. Constructed from the hardest timber available, which at the bow reached four feet in width, the ship was designed to provide maximum warmth and comfort for the crew, who Nansen expected to be drifting with the ice for several years. Launched in October 1892, the ship was named *Fram*.

After leaving Norway in July 1893 and heading northeast into the Kara Sea, *Fram* entered the ice in October. For the first couple of months, the ship made little progress, and indeed, after six weeks the drift had carried it farther south than it had begun. In January 1894, the currents began propelling *Fram* northward, but the pace was achingly slow, and Nansen began to consider the possibility that reaching the Pole might require disembarking and skiing toward it. In March 1895, after *Fram* had passed the previous most northerly latitude ever achieved, Nansen and companion Hjalmar Johansen set out to do just that; after just a couple of weeks of travel, however, Nansen calculated that their progress was not keeping up with their supplies, and the two men instead headed for the Arctic archipelago of Franz Josef Land, which they reached in early August.

Fram, now under the command of Otto Sverdrup, continued its drift until, on August 13, 1896, it finally emerged into open water north of Svalbard. Later that month, it arrived in the Norwegian port of Tromsø, where the crew was swiftly reunited with Nansen and Johansen, who had been rescued from Franz Josef Land by a supply ship.

Although neither Nansen nor *Fram* achieved the Pole, the expedition was immensely successful. In being carried west from the Russian Arctic coast to Svalbard, *Fram* proved the existence of the polar drift that Mohn had first hypothesized and, in the process, further demonstrated that there were no large Arctic landmasses awaiting discovery. Depth soundings taken onboard during the expedition demonstrated that the Arctic was a deep ocean basin—albeit shallower than the world's other oceans. The North Pole, it was now clear, was located neither on land nor on a fixed sheet but amid a constantly swirling mass of pack ice. And *Fram*'s rounded design would inform the construction of future icebreaking vessels, while Nansen's decision to employ a small, trained crew rather than set sail with the bloated companies of expeditions past also proved influential. Among his acolytes was Roald Amundsen, who followed many of his principles while completing the Northwest Passage on *Gjøa*, and who would ultimately use *Fram* as the ship from which he launched his successful assault on the South Pole in 1911.

In September 2019, a century and a quarter after the *Fram* expedition ended, another polar endeavor sought to replicate much of what it had achieved. At 387 feet in length and 82 feet across the beam and with an empty weight of over twelve thousand tons, the *Polarstern* dwarfed Nansen's 127-foot-long, 36-foot-wide vessel; whereas Nansen equipped *Fram* with a 220-horsepower steam engine prior to setting out on his expedition, *Polarstern* boasted four diesel generators with a combined output of almost 19,000 horsepower—which, combined with its double steel hulls, enabled it to break a path through ice floes as much as ten feet thick. Commissioned in 1982 specifically for conducting research in Arctic and Antarctic waters, it had, prior to departure, logged more than 1.7 million nautical miles at both ends of the globe. Among those miles was a 2008 voyage through the Northwest and Northeast Passages in one cruise, making it the first research trip ever to accomplish that goal. Even by its own exalted standards, however, this latest voyage would be challenging; following in the virtual footsteps of Nansen and *Fram*, *Polarstern* would embed itself into the pack ice and drift with it for a year.

The *Polarstern*'s crew enjoyed comforts and advantages that those on the *Fram* could only have dreamed about or barely imagined. Relative to its predecessor, the *Polarstern* was spacious, comfortable, and warm, and such was the ship's power that there was no concern about it being carried helplessly for years on end. Satellite imagery, a global positioning system, and a century and change of further exploration and discovery meant that there was no danger of those onboard not knowing where they were or wondering if they would encounter a previously uncharted landmass. Modern satellite communications enabled the ship and its occupants to remain in regular contact with those ashore. And by constructing runways on the ice, the *Polarstern*'s passengers and crew were able to receive provisions via Twin Otter aircraft, which also enabled onboard personnel to rotate in and out.

That ability to bring on new people and allow existing members to depart did face an unexpected challenge when, partway through its mission, the world the ship had left behind succumbed to the coronavirus pandemic. When one of the aircraft team, who had not yet traveled to the ship, tested positive for COVID-19 shortly after the pandemic's outbreak, resupply flights were immediately canceled, causing some onboard to remain there for two months longer than anticipated and forcing the *Polarstern* to leave the ice for three weeks to rendezvous with two resupply vessels in a Svalbard fjord. In an interesting twist, those onboard as the pandemic broke out were, despite their isolation and their relative privations, among the freest people left on the planet in that they were among the last to be able to enjoy close physical contact with other humans.

In the end, although COVID-19 raised numerous logistic hurdles, the expedition continued largely as planned and concluded when the *Polarstern* returned to port on October 12, 2020, 389 days after departure. Dubbed the Multidisciplinary Drifting Observatory for the Study of Arctic Climate (or MOSAiC), the endeavor ultimately involved more than six hundred active participants onboard and many more trawling through and analyzing the data the expedition collected. The largest scientific expedition ever conducted in the Arctic, its research areas included multiple atmospheric, oceanographic, and ecological studies: from exploring how Arctic sea life survives the long, cold, dark winter to monitoring deep ocean currents and examining the interactions between ocean, ice, and atmosphere. Central to virtually every aspect of the undertaking, however, was the region's sea ice: its formation, its melting, its interactions with the water below and the air above, and the feedbacks that drive its growth and retreat. The motivation for such a mammoth venture needed little explanation: that sea ice is disappearing, more rapidly than anticipated, as the world warms and the Arctic warms fastest of all.

A 2019 analysis of seventeen climate models dating back to the 1970s found that fourteen accurately predicted the response of global temperatures as levels of carbon dioxide increased in the atmosphere. (The estimates of two were too high, and one was too low.) As much as climate change deniers have long viewed models as the soft underbelly of climate science, of being essentially unverifiable and of consistently overstating likely warming, the opposite is true: the 2021 Nobel Prize in Physics was awarded to pioneering climate modelers Syukuro Manabe and Klaus Hasselmann, as well as theoretical physicist Giorgio Parisi, because, said Thors Hans Hansson, chair of the Nobel Committee for Physics, "our knowledge about the climate rests on a solid scientific foundation, based on a rigorous analysis of observations." As far back as 1970, in a seminal paper that was the very first to make a specific projection of future warming, Manabe argued that global temperatures would increase by 0.57°C (1.03°F) between then and the year 2000. The actual recorded warming was a remarkably close 0.54°C (0.97°F).

In fact, climate scientist Michael Mann told me at the time of the Nobel announcement, he and his colleagues "were dismissed as alarmists for the predictions that we made, but the predictions, if anything, turned out to be overly conservative, and we're seeing even greater impacts than we expected to see." One area in which the rate of change has outstripped predictions, and continues to do so, is the Arctic.

Researchers from the University of Gothenburg in Sweden concluded in 2023 that models underestimated both the volume and temperature of Arctic Ocean currents and thus their role in melting sea ice; that same year, a team from the University of New South Wales in Australia argued that models failed to account adequately for stratospheric polar clouds—wispy, rarely visible to the naked eye, unique to the polar regions, and highly efficient trappers of heat.

What is beyond dispute is that not only is the Arctic warming, it is doing so at a substantially faster rate than any other region on Earth. Even here, however, the rate of change appears to have been substantially underestimated. Whereas in 2021 the Intergovernmental Panel on Climate Change (IPCC) offered that the Arctic was warming twice as fast as the rest of the planet, and that year's Arctic Monitoring and Assessment Programme (AMAP) report stated it was three times as fast, a 2022 study by Mika Rantanen of the Finnish Meteorological Institute and colleagues concluded that it was in fact warming fully four times as fast.

Scientists refer to the region's faster and more substantive warming as "Arctic amplification," and there are a number of reasons for its existence. The "weather layer" of the atmosphere known as the troposphere is thinner above the poles than at the equator and so requires less warming; furthermore, in the more humid air of the tropics, a greater proportion of the Sun's energy is expended in evaporation, whereas in the drier Arctic, that energy leads directly to heating. Equatorial regions receive almost constant sunlight, warming the surface and causing warm air to rise into the atmosphere; in the Arctic, however, the more acute angle at which sunlight strikes means that the ground is warmed less directly, causing less warm air to rise and mix with the atmosphere above, ensuring that more warming remains near the surface.

The single biggest factor in accelerated Arctic warming, however, involves what is arguably the region's defining feature: its blanket of snow and ice and in particular the diminution of the sea ice cover that allowed Inuit civilizations and their predecessors to thrive and thwarted the incursions of wave upon wave of European explorers. Not only did sea ice form as a consequence of colder temperatures, it also helped perpetuate those colder temperatures by reflecting sunlight into space, setting in motion a feedback loop that has helped keep the bulk of the Arctic significantly colder than the rest of the world north of Antarctica. However, as that ice begins to melt, it exposes more of the ocean surface beneath; because that ocean is darker, it absorbs the light that the ice would have reflected. So, as Arctic waters become warmer, they lose ice and absorb more sunlight, which makes them warmer still, causing more ice to melt, and so on.

Unsurprisingly, therefore, as the world has warmed and the Arctic especially so, sea ice has diminished, and its retreat has fed further retreat. And as it has withered in extent, so has it declined in age and thickness, with the thickest, oldest ice thinning year after year. Most models predict that eventually, by midcentury, the majority of the Arctic will be largely or even entirely without sea ice for a short period at the end of summer, the bulk of the ice that remains being that which forms each winter only to disappear again the following summer.

———————∞———————

More even than the Northwest and Northeast Passage alternatives, the viability of a Transpolar Sea Route requires not just a significant decrease in sea ice along its route but its virtual disappearance. Even during the scenario just envisioned, in which sea ice in the Arctic Ocean effectively disappears at the end of summer, ice would remain throughout winter, in spring and fall, and in diminished levels over the course of summer—and, most likely, in unpredictable patches at its height. Not until closer to the turn of the century would ice-free waters reach an extent and duration to enable greater predictability for planning shipping routes and a longer duration in which to execute them—although here it is worth recalling the point made a short while earlier that predictions of Arctic change have by and large proven insufficiently pessimistic.

That doesn't necessarily mean that a Transpolar Sea Route—or, if one prefers, Transpolar Passage—is a nonstarter for the next several decades, but it does mean that if it is to become a commercially viable proposition in the interim period, certain realities need to be addressed, and some types of shipping will be more likely than others. Even as Arctic sea ice retreats in extent, it does not do so in a linear fashion: for the first twenty-two years of the satellite record, from 1979 to 2001, the trend was clearly one of decline, but low summer ice years were generally followed by a regrowth to typical levels over the winter. But in 2002, and then again in 2005, summer sea ice extent fell to new lows, of 2.3 million and 2.15 million square miles, respectively, and on each occasion winter ice formation struggled to keep up. Then, in 2007, came a precipitous collapse, to a mere 1.65 million square miles, fully two Texases below the previous low. That marked the beginning of a "new normal" in which sea ice minimum has not reached 2 million square miles since; even so, at the time I write this, in July 2024, only two summers (2012 and 2020) have experienced lower lows than 2007, although three—2016, 2018, and 2023—effectively equaled it.

All of which is to underline that change in Arctic sea ice, while underway and profound, is not predictable with precision—at least, not the degree of precision that industries such as commercial shipping generally crave. It is possible, perhaps probable, that any Transpolar Passage would see little to no commercial activity until such time as it is predictably clear of all ice cover for a period of weeks to months each summer, in the back half of the 2000s. But let us assume that at least some companies and countries decide to steal a jump on their rivals, to probe the boundaries of what is possible in the Arctic and to seek to push back on the temporal and spatial limitations that the region's ice cover currently presents and is likely to for the near future.

Doing so would present challenges beyond those of shipping via the Northern Sea Route or even the Northwest Passage. For one, unlike either alternative, it would, at least initially, rise and fall almost entirely on its functionality as a route for transit shipping—that is, vessels that leave a location in, say, northern Europe carrying goods to, for example, South Korea. Whereas the Northern Sea Route, for all Vladimir Putin's boosterism, is presently largely functioning as an internal waterway, that would not be an option for a transpolar passage, which would have no local communities to serve en route, save perhaps those in the Bering or Fram Strait at either end of the journey. Nor, again unlike the Northern Sea Route, are there any extractive or manufacturing industries—no mines, no mills, no factories—for which to provide supplies or from which to distribute goods. As a consequence, any shipping company considering using the route would be placing all its eggs in a very-long-haul basket, with few guarantees of an unobstructed pathway outside of that short peak-summer window.

These initial stages would most likely not be especially conducive to the bulk of seaborne goods commerce, which is conducted on a just-in-time basis by what is known as liner shipping: ships following regular routes on specific timetables, offloading goods at ports from which they are then conveyed inland, the whole process conducted with a timeliness that allows businesses to receive goods and parts as they need them, relieving them of the pressure of having to maintain high levels of inventory. It's a system that is both extremely efficient and highly vulnerable to disruptions—hence the significant supply chain interruptions when the *Ever Given* became stranded in the Suez Canal, or the still-reverberating shocks to the system caused by the COVID-19 pandemic. In time, if and when the ice cover of the Arctic Ocean basin has retreated completely for a period of several months or more, a Transpolar Sea Route may be a functional element of that system, but as long as it remains a somewhat mercurial pathway with more potential than predictability, any commercial traffic that it sees is more likely to be bulk carriers transporting less time-sensitive goods—coal, steel, cement, petroleum, liquid natural gas, cooking oil—and transit shipments between the Atlantic and Pacific that require somewhat rapid delivery but are unsuitable for air transport: specifically, large items such as automobiles.

In a chapter in a book edited by Kristina Spohr and Daniel Hamilton, *The Arctic and World Order*, Mia Bennett and colleagues note that, outside of ice-free periods in summer, both maritime safety and the demands of insurance companies would require that transits be conducted either by ice-strengthened ships or in the company of icebreakers. The latter is, as we have seen, a requirement for vessels navigating the Northern Sea Route and is a significant cost outlay. Furthermore, while Russia subsidizes the construction and operation of icebreakers in the Northern Sea Route for its own purposes and to entice the passage's use, that role would presumably go unfulfilled in the international waters of the Transpolar Passage, further increasing expenditure. At the same time, thinning ice in the Central Arctic Ocean could well enable ships one step down from icebreakers—Polar Class vessels, which, in their sturdiest configuration, are able to crunch through floes up to four meters thick—to shoulder much of the burden of any transpolar trade; however, there is considerable cost to building such vessels, which raises issues of economic efficiency if they are to spend part or a large part of their journey in relatively open water. Similar considerations come into play when contemplating the viability of double-action vessels such as those used in the Yamal LNG project: ships that steam forward through open water and turn around to grind astern through ice.

All of which suggests, wrote Bennett and her coauthors, that any transpolar shipping industry is likely to utilize a hub-and-spoke system wherein standard open-water ships carry goods to a port just south of the Arctic Ocean, where their cargo is switched to Polar Class vessels for the journey across the top of the world before being transferred anew on the other side for the final legs.

The two possible entrances to a Transpolar Sea Route are via the Fram Strait—the passage of water between Greenland and Svalbard through which Nansen's ship emerged from its induced imprisonment—on the Atlantic side and the Bering Strait on the other, so a search for appropriate transshipment locations leads us to those areas, where we find that plans for such eventualities are at least in the discussion stage at several possible venues.

Longyearbyen, the principal settlement in Svalbard, presents itself as a clear candidate on the Atlantic side. As ice has retreated and tourism has expanded, port calls have already increased approximately tenfold since the beginning of the century, prompting the creation of one new floating and three permanent quays to accommodate ships up to a thousand feet in length. Bennett and colleagues wrote that the port's draft still isn't sufficient for some of the largest container ships but that Norway, which governs Svalbard, has allocated resources for further port expansion.

Despite being nearly a thousand miles farther south, Iceland too has been eyeing the opportunities presented by a potential Transpolar Sea Route, with the government looking to develop a deepwater port in Finnafjord, in the country's northeast, which might hold a logistic and competitive advantage should the transpolar passage become a regular route for the largest container ships. In 2019, the German company Bremenports signed an agreement with the government in Reykjavík to build and operate the port through 2040.

In the Bering Strait, options exist on both the Russian and American sides. The Russian port of Providentiya would seem to have an inherent advantage in that it already serves as the de facto eastern gateway for the Northern Sea Route, is deeper than most alternatives, and already possesses infrastructure such as an oil spill response system. The United States, meanwhile, appears to be zeroing in on Nome as its alternative; it already acts as a transshipment hub for western Alaska and as a staging point for voyages farther north, and in 2020 the US Army Corps of Engineers approved a plan to dredge the harbor to increase its depth and suitability for larger ships.

For now, the uncertainty surrounding the Transpolar Passage and the fact that any regular shipping is unlikely to come to fruition before midcentury means that most countries' commitments to developing the necessary infrastructure or investing in suitable fleets is presently tentative. Some actors, however, appear to be showing more concerted interest in the prospect of sailing across the top of the world, including those who would not normally be remotely associated with the Arctic.

In 2012, for example, a 21,000-ton icebreaker left port on July 2, sailed north through the Bering Strait and then west along the Northeast Passage to the Barents Sea, arriving in Iceland in August. From there, it set out via the Fram Strait for the North Pole and, despite falling short of that goal, adjusted course and traversed the Arctic Ocean, sailed south back through the Bering Strait, and arrived in Shanghai on September 27.

The name of the ship was *Xue Long*, or *Snow Dragon*, and it was Chinese.

CHAPTER 12

Building a Polar Silk Road?

In her book *China as a Polar Great Power*, Anne-Marie Brady recounts the scene on September 2, 2015, when, for the first time, the Pentagon reported the visit of five vessels—three warships, an amphibious vessel, and a supply ship—from the People's Liberation Army Navy (PLAN) into US territorial waters off Alaska. The Pentagon's response, she wrote, was "muted": the vessels were exercising their right to innocent passage through US waters, and the United States does much the same in the Taiwan Strait and the South China Sea, among other locales.

Nonetheless, neither the timing nor the location of the vessels' appearance was coincidental. The ships showed up the day before China's biggest-ever military parade and just as President Barack Obama was closing out a trip of his own to Alaska, during which he touted his administration's Arctic policies. Nor were the ships merely passing through: they had been participating in a military exercise with Russian forces near Vladivostok, a Russian naval port close to the border with North Korea, and had steamed directly north into the Bering Sea, transiting the Aleutian Islands; then, their presence having been noted, they turned around and steamed south back to China.

As Brady noted, the move had its desired effect. News of the vessels' presence "drowned out Obama's visit to the Arctic" as commentators noted China's ability to project power even to the northernmost reaches of the planet.

So, what is the extent of China's interest in the Arctic? Given that the country is neither adjacent to nor possessed of a long history of involvement in the region, what is its motivation? The answer to the second question is both complex and straightforward: it is at once neither geopolitical, commercial, nor military, and yet simultaneously all of these: an important component of China's progression toward genuine superpower status, a forward-looking approach to the challenges and opportunities presented by a changing Arctic, and a move to secure a prominent seat at the table to influence and benefit from decisions that inevitably will be made in response to those Arctic changes.

As with many such official documents, "China's Arctic Policy," a white paper published by the State Council Information Office of the People's Republic of China in January 2018, manages to say both quite a lot and very little. It characterizes Beijing's policy goals in the Arctic as "to understand, protect, develop and participate in the governance of the Arctic, so as to safeguard the common interests of all countries and the international community in the Arctic, and promote sustainable development of the Arctic"—the kind of pablum that would be equally at home in similar documents emanating from London or Washington, DC. It summarizes the country's basic principles concerning its Arctic operations with such corporate buzzwords as "respect, cooperation, win-win result and sustainability."

Dig a little deeper and read between the lines and you'll find some bread-crumbs, however. In a section on international shipping routes, it states:

> As a result of global warming, the Arctic shipping routes are likely to become important transport routes for international trade. China respects the legislative, enforcement and adjudicatory powers of the Arctic States in the waters subject to their jurisdiction. China maintains that the management of the Arctic shipping routes should be conducted in accordance with treaties including the UNCLOS and general international law and that the freedom of navigation enjoyed by all countries in accordance with the law and their rights to use the Arctic shipping routes should be ensured. China maintains that disputes over the Arctic shipping routes should be properly settled in accordance with international law.

The third sentence, concerning the management of shipping routes and international law, is a subtle nod of agreement with the United States' position on the Northwest Passage: that it is an international waterway through which countries should be able to exercise free passage. The same presumably holds for the Northeast Passage, notwithstanding China's more cooperative working relationship with Russia along the Northern Sea Route. The previous sentence, acknowledging the existing powers of Arctic states, displays a hint of the passive-aggressive: "the waters subject to their jurisdiction" is doing a lot of heavy lifting, and left unsaid is the issue of exactly which and how extensive those waters might be.

There is another sentence in the same section of relevance to the themes discussed in this book. "China hopes to work with all parties . . . through developing the Arctic shipping routes," it states. "It encourages its enterprises to participate in the infrastructure construction for these routes and conduct commercial trial voyages in accordance with the law to pave the way for their commercial and regularized operation."

Three years later, in 2021, the National People's Congress formally adopted its latest Five-Year Plan, the fourteenth in total and the first to mention the Arctic. That mention was blink-and-you'll-miss-it brief, but what it chose to include could reasonably be inferred to highlight Beijing's priority in the northern polar regions. China will, it read, "participate in practical cooperation in the Arctic and build the 'Polar Silk Road.'"

————— ∞ —————

That China should be invested in the idea of shipping routes through the Arctic to the point of wanting to develop infrastructure to support those routes and even give the concept some catchy branding should, on one level, be no surprise. China boasts, by at least one measure, the largest merchant fleet in the world; by other measures, it lies second behind Greece, but whatever metric one chooses—be it gross tonnage, deadweight, or number of vessels—its position on the charts has been rising, and having overtaken Japan's fleet in 2018, it is at the very least competitive in size with the Greeks' and increasing at a faster rate.

In such circumstances, it would arguably be a dereliction of duty not to read the tea leaves and take as many necessary precautions as possible for a world in which shipping routes through the Arctic are viable and competitive. The notion of a Polar Silk Road, however, suggested something deeper and more substantive.

But exactly what it means and how much import Beijing assigns to it is unclear. The Polar Silk Road was announced as a component of China's Belt and Road Initiative, which has been described as "one of the most ambitious infrastructure projects ever created." When it was announced in 2013, it was portrayed as a "vast collection of development and investment initiatives . . . originally devised to link East Asia and Europe through physical infrastructure," which in the decade since "has expanded to Africa, Oceania, and Latin America, significantly broadening China's economic and political influence." It began as two separate projects—the overland Silk Road Economic Belt and the Maritime Silk Road—that were designed to evoke the original Silk

Road, which arose during the Han dynasty, an era lasting approximately four centuries that began in 206 BCE and reached a peak during the first millennium CE. The Silk Road brought a thriving trade to the area now occupied by the central Asian nations of Afghanistan, Kazakhstan, Kyrgyzstan, Tajikistan, Turkmenistan, and Uzbekistan, as well as to India and Pakistan to the south, and saw "valuable Chinese silk, spices, jade, and other goods [move] west while China received gold and other precious metals, ivory, and glass products."

According to an analysis by James McBride, Noah Berman, and Andrew Chatzky of the Council on Foreign Relations, Chinese president Xi Jinping's vision of the Silk Road Economic Belt "included creating a vast network of railways, energy pipelines, highways, and streamlined border crossings, both westward—through the mountainous former Soviet republics—and southward, to Pakistan, India, and the rest of Southeast Asia. Such a network would expand the international use of Chinese currency, the renminbi, and 'break the bottleneck in Asian connectivity.'" As for the Maritime Silk Road: "To accommodate expanding maritime trade traffic, China would invest in port development along the Indian Ocean, from Southeast Asia all the way to East Africa and parts of Europe."

While nominally marketed as a massive global infrastructure project, the entire Belt and Road Initiative also has national and geopolitical aspirations: to bring greater investment and opportunity to its oft-neglected western regions, to restructure and invigorate the Chinese economy more broadly, and to develop trade linkages and diplomatic leverage in the region. Its sheer scope has caused some anxiety in Western capitals, and the inclusion of the Arctic in the maritime component has added to the notion that the world's northernmost region is shaping up to be a geopolitical battleground.

Speaking at the Arctic Circle Assembly in Reykjavík in 2022, Admiral Rob Bauer, chair of the Military Committee of the North Atlantic Treaty Organization (NATO), stated that China is expanding its presence in the Arctic "by leveraging new opportunities provided by the melting ice in the region, including the PSR linking China and Europe, by investing tens of billions of dollars in energy infrastructure and research projects in the region." However, argued Erdem Lamazhapov, Iselin Stensdal, and Gørild Heggelund in an analysis for the Arctic Institute, there is no evidence in support of that contention. They noted that most of the investment China has been making in the Arctic has been in relation to the Northern Sea Route—which, as we have seen, has declined precipitously since Russia's invasion of Ukraine.

Their view is supported by an analysis by Doug Irving for the RAND Corporation that examined the level of Chinese activities in the Arctic and found that, particularly in the North American Arctic (which includes Greenland as well as Alaska and Canada), "there's not a ton going on." Furthermore, whereas Beijing's approach to gaining partnerships for its Silk Road Economic Belt has been largely to buy friends, Arctic nations have been far more circumspect about allowing China a foothold. The RAND review noted that "a Chinese company tried to buy a shuttered U.S. Navy base in Greenland, but the Danish government quashed the idea. . . . Canada blocked a $150 million gold mine deal that would have put Chinese interests too close to military installations. Greenland has held up plans for another Chinese mine over concerns about pollution."

A review of Chinese ambitions for the Arctic by Marc Lanteigne for *The Diplomat* observed:

> During the past five years, many original components of the PSR across the Arctic Ocean have been delayed or scrapped entirely, due either to shifting political winds or specific concerns by Arctic governments relating to the financial and security risks of Chinese investments. The most prominent examples of China-backed initiatives in Arctic and Arctic-adjacent regions that failed to materialize include a railway connection between northern Finland and Norway; a uranium and rare earths mining site at Kuannersuit, Greenland (as well as a long-planned iron mine on the island); liquified natural gas investment in Alaska; land acquisitions in Iceland and Norway; a gold mine purchase in Nunavut; an underwater Arctic communications conduit along the Northern Sea Route (NSR) between Asia and Europe; and a tunnel connecting Estonia and Finland.

Additionally, much has been made of China's plans for increased ice-breaking capacity, and justifiably so: at time of writing, construction is underway on the country's third and fourth icebreakers—*Tan Suo San Hao* and *Ji Di*—which, in addition to *Xue Long* and *Xue Long 2*, would give China, which isn't an Arctic nation, twice as many icebreakers as are presently owned and operated by the United States, which is. But previous talk of China's fleet being top-of-the-line nuclear-powered vessels appears to have evaporated, perhaps as a result of Chinese skittishness at partnering with Russia, which has the most extensive expertise in building nuclear-powered icebreakers.

And while it is notable that the Arctic and the Polar Silk Road were mentioned in a Five-Year Plan, it is perhaps equally significant that the mention was so brief, included only as a passing element in a chapter discussing how, as Trym Eiterjord wrote for the Arctic Institute, China planned to "actively expand the developmental space of the marine economy."

But while it is possible to overstate Chinese intentions in the Arctic, and certainly to become overly fixated on the extent to which they conjure up images of great power conflict at the top of the world, it would equally be a mistake to diminish them. Any perceived quieting of Beijing's interest in the region can be ascribed, argue the authors of the Arctic Institute review, to China pursuing an approach of "crossing the river by touching the stones"—of adapting to circumstances. Many of Beijing's early Arctic ventures have been conducted in conjunction with Moscow; with Russia's invasion of Ukraine making that more difficult—and indeed posing a challenge to all manner of cooperation across the Arctic—China may choose to deprioritize the Polar Silk Road and related activities until a more suitable time. But that does not mean its interest in the Arctic isn't genuine.

China has declared itself a near-Arctic state—a designation of its own invention—and since 2013 has been an observer at the Arctic Council. In addition to contemplating shipping networks through previously ice-choked regions, it has, noted the RAND review, "dispatched research expeditions [and] sought to establish mining and gas operations. . . . It describes itself as an 'active participant, builder, and contributor in Arctic affairs,' one that has 'spared no efforts to contribute its wisdom to the development of the Arctic region.'"

Furthermore, note seasoned Sinologists, China's goals regarding the Arctic are sometimes expressed more forcefully in Mandarin for domestic consumption than in English for a global audience. As noted by Rush Doshi, Alexis Dale-Huang, and Gaoqi Zhang of the Brookings Institution, Xi has, for example, frequently expressed his desire at home to make China a "polar great power," a phrase generally missing from externally facing materials. Similarly, while documents and speeches aimed at the outside world downplay any Chinese thoughts of military competition in the Arctic, internal texts are more explicit about the feeling in Beijing that "the game of great powers" will "increasingly focus on the struggle over and control of global public spaces," including the polar regions. China "cannot rule out the possibility of using force" in this coming "scramble for new strategic spaces."

While, from a Western perspective, the notion of Chinese involvement in the Arctic and particularly the specter, however remote, of military confrontation there may raise hackles, China's motivation does not appear too dissimilar to that of any other country with interests in the Arctic in a warming world. Those motivations include the economic and geopolitical security and advantages offered by the prospect of shipping lanes in the region. But as much as on paths through the ice, at least some countries appear to be keeping a keen eye on what lies beneath it.

———————

On July 22, 2007, the Russian research ship *Akademik Fedorov* left Murmansk, carrying two deep submergence vehicles (DSVs) and accompanied by the nuclear icebreaker *Rossiya*, and set sail for the North Pole. On August 2, after a delay caused by an electrical fault and following some test dives of the DSVs just north of Franz Josef Land, the *Akademik Fedorov* positioned itself over the Pole and launched the submersibles twenty minutes apart. Over the course of a little under three hours, the two subs descended to the seafloor, where they stayed roughly an hour before beginning their ascent. The *Fedorov* returned to Franz Josef Land, resupplied a French scientific expedition, and established a drifting research station on a massive ice floe before heading back to Murmansk and thence Saint Petersburg.

It was a successful outcome to a logistically challenging expedition, including the first ever crewed dive to the seabed at the North Pole, but what caught global attention was what the expedition left behind: a time capsule, containing a flag from Vladimir Putin's United Russia party; and a one-meter-tall titanium Russian flag.

The latter in particular provoked reactions ranging from stern and studied to borderline apoplectic. A US Department of State spokesperson, Tom Casey, stated that planting the flag "doesn't have any legal standing or effect" on any territorial claims on the North Pole region. Canadian foreign minister Peter MacKay was more direct, fulminating, "This isn't the 15th century. You can't go around the world and just plant flags and say, 'We're claiming this territory.'"

In response, Moscow's perpetually phlegmatic foreign minister Sergey Lavrov professed himself "amazed" by MacKay's comments, protesting innocently, "We're not throwing flags around. We just do what other discoverers did. The purpose of the expedition is not to stake whatever rights of Russia, but to prove that our shelf extends to the North Pole. By the way on the Moon it was the same."

Lavrov's "We're not staking a claim, just asserting that our claim is valid" defense referred to a formal submission that Moscow made in December 2001 to the United Nations Commission on the Limits of the Continental Shelf, arguing that Russia's shelf extended northward beyond the two-hundred-mile exclusive economic zone established under the United Nations Convention on the Law of the Sea (UNCLOS), deep into the Arctic Ocean. Russia's rationalization centered on the Lomonosov Ridge, a mountain range that extends from Ellesmere Island to a point near the North Pole and onward to a point near the New Siberian Islands off Russia's Arctic coast, effectively bifurcating the Arctic Ocean seafloor into Eurasian and Amerasian Basins. Its existence was unknown until Soviet researchers, working from a camp on the sea ice, detected unexpectedly shallow waters off the New Siberian islands. Six years later, they published a map showing a mountain range, which they named after eighteenth-century naturalist Mikhail Lomonosov, one of Russia's most celebrated scientists, who in 1763 was the first to create a map of the Arctic Ocean that did not include any islands or large landmass in its center.

In his book *Cold Rush: The Astonishing True Story of the New Quest for the Polar North*, Martin Breum wrote that "Lomonosov created the foundation for a wave of Russian activism. In 1926, the Soviet leaders sent out a decree saying that any rock in the Arctic Ocean, from the coast of Siberia to the North Pole itself, was to be considered Soviet territory. Ivan Dmitrievich Papanin became a national hero in Russia in 1937 when he, with a team of scientist colleagues, flew to the North Pole as the start of a spectacular science expedition. With modern navigation equipment, Russia became the first country to prove that its envoys had been at the North Pole—on a mission directly encouraged by Stalin."

When Russia submitted its claim to the UN continental shelf commission, the commission neither accepted nor rejected the submission, instead suggesting the need for further research. Following the 2007 expedition, Russia's Ministry of Natural Resources and Environment released a statement that, lo and behold, analysis of the results confirmed that "the Lomonosov Ridge . . . is therefore part of the Russian Federation's adjacent continental shelf." According to an email from the US embassy in Copenhagen, uncovered by WikiLeaks in 2011, the submersible dive that planted the Russian flag was conducted at the behest of Putin's United Russia party.

———————⊷———————

Twenty years earlier, another leader in Moscow advocated an entirely different kind of policy in the Arctic. On October 1, 1987, Mikhail Gorbachev gave a speech that became known, after the city in which he delivered it, as the Murmansk Initiative. It set out a series of unifying principles he believed should govern the approach of the Soviet Union, and indeed the global community, toward the northern polar regions.

Some of what the final Soviet leader said would not feel out of place today: he called, for example, for the opening of the Northern Sea Route to international shipping, "with ourselves providing the services of ice-breakers." Other proposals seem simultaneously very much of their time and yet also of renewed relevance: for example, that the Arctic should be a nuclear-free zone, with Moscow prepared to act as a "guarantor" and offering the prospect of removing submarines equipped with ballistic missiles from the Soviet Baltic Fleet. Other items proved to be genuinely revolutionary in terms of Arctic governance: from cooperating in environmental protection and management to organizing an international scientific conference that might lead to the establishment of an Arctic Research Council.

These were heady times; during the short window between the introduction of glasnost and perestroika and the collapse of the Soviet bloc, almost anything seemed possible in terms of international relations, and the Murmansk Initiative contributed to that giddiness. Beyond his specific policy proposals, what was notable was Gorbachev's overarching vision of the region: "The Arctic is not just the Arctic Ocean but also the northern tips of three continents—Europe, Asia, and America. It is the place where the Euro-Asian, North American and Asia-Pacific regions meet, where the frontiers come close to one another and the interests of states belonging to mutually opposed military blocs and nonaligned ones cross."

The government of Finland accepted the challenges implicit in Gorbachev's speech, issuing a call for ministers of the eight Arctic nations to meet in the town of Rovaniemi. There, on June 14, 1991, they issued the Rovaniemi Declaration and established the Arctic Environmental Protection Strategy (AEPS). The AEPS sought to address a wide range of environmental issues on land and sea in the Arctic and recognized the importance of addressing those issues—notably pollution—for the well-being not only of the region's environment but also of the humans who lived in and were a part of that environment. To that end, the AEPS invited three Indigenous organizations—the Inuit Circumpolar Council (ICC), the Saami Council, and the Russian Association of Indigenous Peoples of the North (RAIPON)—to be permanent participants. Three more ministerial meetings of the AEPS followed in 1993, 1996, and 1997; but during those discussions Canada in particular began arguing for more formal arrangements to strengthen international cooperation and promote environmental protection and sustainable development in the region. The result was the 1996 Declaration on the Establishment of the Arctic Council (Ottawa Declaration), which was followed the succeeding year by the Alta Declaration (it will not have escaped notice that declarations are de rigueur in Arctic geopolitics), which folded the AEPS into the Arctic Council's affairs. The council operates several working groups of import to Arctic governance, including several it inherited from the AEPS, among them the Arctic Climate Impact Assessment, the Arctic Human Development Report, the Arctic Resilience Report, and the Arctic Marine Shipping Assessment; but it has no authority to oversee or rule on territorial disputes. However, it does provide a regular forum for cooperation, coordination, and interaction among the eight Arctic states, observer nations, and Indigenous Permanent Participants, under which it can develop legally binding treaties. It also enables members, observers, and participants to work together on the implementation of international treaties and organizations that are binding on certain activities in the Arctic, such as the Law of the Sea or the regulations of the International Maritime Organization.

In other words, despite the absence of a singular convention like the Antarctic Treaty, which regulates human activities at the other end of the world, Arctic nations consider their region to be subject to effective governance; and so, when parts of the global community responded to the Russian flag planting at the North Pole by fretting about a geopolitical vacuum at the top of the world and the arrival of a new, and this time literally, cold war, an unusual thing happened. The Arctic states that on paper had the most reason to be offended or threatened by the Russian action and to encourage the establishment of some form of legal safeguard to protect their interests in the face of Moscow's apparent territorial adventurism in fact joined hands with Russia against the outside world. Reasoning that working out their differences among themselves was a better option than allowing outside interests—such as China—to become embroiled and invested in any disputes, representatives of the five coastal Arctic states (the Arctic Council members minus Iceland, Finland, and Sweden, known as the Arctic Five) convened the Arctic Ocean Conference in Ilulissat, Greenland, in May 2008, at which they issued—naturally—the Ilulissat Declaration, the subtext of which was essentially "Yes, we know the Arctic is undergoing convulsive change, but trust us, we've got this."

The declaration began by recognizing that the Arctic Ocean "stands at the threshold of significant changes. Climate change and the melting of ice have a potential impact on vulnerable ecosystems, the livelihoods of local inhabitants and indigenous communities, and the potential exploitation of natural resources." After outlining their combined sovereign and jurisdictional rights in the region, and their commitment to existing international frameworks such as that provided by the Law of the Sea, the Arctic Five stated, "We therefore see no need to develop a new comprehensive international legal regime to govern the Arctic Ocean. We will keep abreast of the developments in the Arctic Ocean and continue to implement appropriate measures." They concluded with a reassurance that they were fully aware of the unfolding issues, thank you very much, and were already discussing the best ways to manage them:

> The five coastal states currently cooperate closely in the Arctic Ocean with each other and with other interested parties. This cooperation includes the collection of scientific data concerning the continental shelf, the protection of the marine environment and other scientific research. We will work to strengthen this cooperation, which is based on mutual trust and transparency, inter alia, through timely exchange of data and analyses.
>
> The Arctic Council and other international fora, including the Barents EuroArctic Council, have already taken important steps on specific issues, for example with regard to safety of navigation, search and rescue, environmental monitoring and disaster response and scientific cooperation, which are relevant also to the Arctic Ocean. The five coastal states of the Arctic Ocean will continue to contribute actively to the work of the Arctic Council and other relevant international fora.

The declaration made clear that Arctic nations felt far better equipped to handle disputes among themselves than to face the prospect of the region becoming a true global commons.

True to their word, the Arctic nations have indeed worked together diligently and effectively, through the Arctic Council and in cooperation with other organizations, on a number of issues in the years since. On three occasions, they developed legally binding treaties under the council's auspices—on aeronautical and maritime search and rescue in the region (in 2011), on marine oil pollution preparedness and response (2013), and, fulfilling Gorbachev's vision, enhancing international Arctic scientific cooperation (agreed in 2017). All Arctic nations are bound by the provisions of the International Maritime Organization's Polar Code, agreed in 2017, which imposes structural, safety, operational, and environmental standards on shipping in the polar regions. And in 2018, the Arctic Five joined with China, Japan, the Republic of Korea, and the European Union to agree on the Central Arctic Ocean Fisheries Agreement, which seeks to prevent overfishing in the high seas area of the Arctic Ocean (i.e., the area surrounding the North Pole as far south as Arctic coastal states' two-hundred-mile exclusive economic zones) as part of a "long-term strategy to safeguard healthy marine ecosystems" as ice retreats and accessibility for fishing vessels increases.

Such shows of solidarity did not, however, mean that those disputes disappeared.

Key to determining whether an underwater ridge is indeed an extension of a country's continental shelf is establishing that it is a piece of submerged land and not merely a geological formation that arose independently—by, for example, seafloor spreading, the phenomenon responsible for the Mid-Atlantic Ridge, which stretches from Iceland to Antarctica. A 2012 analysis by a team from the Geological Survey of Denmark and Greenland (GEUS) sampled rocks from the ridge and concluded that it was indeed submerged land, the result of a geological process that took place 470 million years ago. Subsequent studies have added support to that claim but not necessarily to Russia's assertion of sovereignty. From a Russian perspective, the ridge may extend from its continental shelf and reach out toward North America, but from the viewpoint of Denmark and Greenland, it travels in the opposite direction. Indeed, Christian Knudsen, a member of the GEUS team, told the BBC's Martha Henriques in 2020, "It's definitely continent. And it's continent that is similar to what we find in eastern Greenland—it is a continuation of Greenland, that's our main point."

Sure enough, in December 2014, Denmark submitted its own claim to an area of almost 350,000 square miles that extended from the Greenland coast past the North Pole and to the edge of Russia's exclusive economic zone—a larger and more assertive claim than that filed by Moscow.

The northwest Greenland coast is an almost literal stone's throw away from Ellesmere Island; accordingly, Canada subsequently filed a claim of its own—two, in fact: one in 2019 that encompassed an area of more than 450,000 square miles and a revised, expanded follow-up in 2022 that added a further 230,000 square miles.

Those claims followed one by Norway, made the year before the Russian expedition to the North Pole, which did not end up significantly overlapping with any of the other nations' but sought to clarify the legal situation around three specific areas. Following feedback from the UN continental shelf commission, it set the northern boundary of its claim roughly 350 miles short of the North Pole.

Not to be outdone, the United States finally joined the club in late 2023, lodging a claim for an area of approximately 620,000 square miles in the northern Beaufort Sea, which skirted Russian claims but overlapped with portions of Canada's. It left less than 1 percent of the Arctic Ocean seabed unclaimed and left the claimant nations at something of a stalemate.

The US claim came with a wrinkle in that Washington has yet to ratify UNCLOS, the legal authority for such matters, under which the commission on continental shelf limits was established. Unsurprisingly, news of it being lodged was not exactly applauded in Moscow.

"The unilateral expansion of borders in the Arctic is unacceptable and can only lead to increased tensions," Nikolai Kharitonov, a lawmaker who heads the Russian parliament's Arctic committee, told RIA Novosti, as Mike Eckel, Wojtek Grojec, and Ivan Gutterman reported for *Radio Free Europe / Radio Liberty*. "Before anything, it's necessary to prove the geological affiliation of these territories, as Russia did in its own time."

It isn't difficult to find a rationale for such relatively sudden and determined registration of interests in the Arctic Ocean. It is certainly no accident that they have sprung up following and in response to Russia's 2007 expedition and the Ilulissat Declaration, in that they are ripostes not just to Moscow but also to Beijing and any other capitals that may have designs on operating within the Arctic realm. There is also the logistic and practical consideration of countries (the United States excepted) not lodging claims until they have ratified the Law of the Sea—which was in 1997 for Russia, 2003 for Canada, and 2004 for Denmark. But the reason *super omnia* for countries registering an interest in the seabed of the Arctic Ocean is, of course, the realization that the seabed might soon become accessible for the first time in human history—and, further, the possibility that it might prove very much worth accessing.

Mining the Arctic, literally and figuratively, is nothing new. Longyearbyen, capital of the Norwegian Arctic archipelago of Svalbard (formerly Spitsbergen), was founded in 1906 solely to extract coal from the ground. The first oil began pumping from Alaska's north coast in the 1960s. The waters of Svalbard, Greenland, and other parts of the Eastern Arctic were the scene of centuries of intensive commercial whaling beginning in the early 1600s.

The potential availability of several million square miles of seabed, however, would seem on the face of it to be a potential game changer. According to a 2012 analysis by the US Geological Survey, the Arctic may contain as much as 13 percent of undiscovered oil reserves and as much as 30 percent of undiscovered natural gas resources. That would add up to approximately 90 billion barrels of oil—almost four times as much as the oil in the Prudhoe Bay field that kick-started oil drilling in Arctic Alaska.

That may seem to be 90 billion reasons for countries to begin jostling for position and arguing for territory, 90 billion reasons to assume the Arctic is indeed about to become a new ground zero for superpower and regional power jousting. But those figures come with a caveat: the overwhelming majority of those reserves are beneath existing continental shelf claims, within Arctic nations' two-hundred-mile limits. There is little of consequence to be added via new territorial claims that encompass the North Pole.

So, why are the Arctic Five lodging such claims? The simple answer is that they can. It's easy; even pretending to be committed to fighting global warming by agreeing on the wording of end-of-climate-conference memoranda is difficult; and it is far, far harder to actually take the necessary steps. In contrast, it is relatively simple to file a territorial claim, particularly if the scientists and lawyers you will be using to examine, define, and defend that claim are from government agencies and thus already on payroll. Countries may be intrinsically resistant to committing to any kind of self-restrictions for such paltry purposes as saving the planet, but they are on the whole entirely comfortable with expansion. The entire history of the Western world is of nation-states seeking to expand their territories, to extend their boundaries and influence. It's what they do. It's in their DNA.

Recognizing that sea ice is disappearing and seeing an opportunity to claim the ocean floor that is now exposed? That's easy; the system is literally set up to do just that. Tackling the economies, industries, and sociologies inherent in causing that sea ice to disappear in the first place? That involves attacking the fundamental pillars of the system itself, and that's hard.

Nevertheless, there is, depending on one's perspective, something either deeply meta or inherently offensive about the notion that the immediate response of governments to the disappearance of sea ice is not to take the necessary, if difficult, steps required to reduce and reverse that disappearance, but instead to take advantage of it in case it allows them to extract more of the fossil fuels that caused sea ice to begin retreating in the first place.

But then, given the repeated defeats, humiliations, and horrors that sea ice has visited on the individual and collective European and American experience and imagination, one could be forgiven for assuming that a world without sea ice is a world for which plenty have yearned all along.

CHAPTER 13

The Meaning of Ice

The city of Marseille is one of the major ports in the Mediterranean region, an "exuberantly multicultural" melting pot, a gateway to the nation beyond. With a population of just under nine hundred thousand and an area of 93 square miles (241 square kilometers), it is the second most expansive city in France; and, at 2,600 years of age, it is widely considered the oldest. It was founded, as Massalia (sometimes known as Massilia), in approximately 600 BCE by settlers from the eastern Greek town of Phocaea—people who were, wrote historian and author Barry Cunliffe in *The Extraordinary Voyage of Pytheas the Greek*, "among the more entrepreneurial of the Ancient Greeks, renowned, according to the historian Herodotus, for sailing in warships called penteconters to explore the Aegean and the northern part of the western Mediterranean."

In combination with other settlements the Phocaeans established to east and west, Massalia would, like its modern-day equivalent, have seen huge amounts of trade from the Mediterranean to the hinterland and back again. Indeed, it has been claimed that the Massaliots were the unchallenged masters of trade of the time, a status underlined by the presence of Greek and Greek-themed coinage that circulated among Gaulish and Celtic tribes across much of modern-day France and even as far as parts of Britain. From the remnants of amphorae in ruins of the time, we know that wine was among the commodities that flowed north; what made the return journey is largely speculative but included possibly slaves; probably furs, silver, and copper; and almost certainly tin, an essential component of bronze and thus in high demand but also in low supply in ancient Greece.

In approximately 500–250 BCE, Massalia was caught up in an ongoing struggle for supremacy between Greece and Carthage, which at various times and for lengthy stretches saw Greek access to parts of the western Mediterranean greatly restricted. As a result, traditional markets for tin and other commodities were closed off for long periods, forcing the Massalians to look elsewhere. One man looked farther and wider than anyone else. His name was Pytheas, and he was an astronomer and geographer of some repute. Eloise McCaskill, a mid-twentieth-century Arctic scholar, noted:

> From the little that is known of him he still stands in history as one of the world's greatest geographers. For it was he who first marked places on the earth by dependable signs from the heavens. . . .
>
> He could build scientific instruments from the description of others and devise new instruments of precision. He became the first in known history to measure accurately the distance of a place from the equator; he fixed the latitude of Massilia by instruments of which he was the inventor. He was the first among the Greeks to arrive at any correct notion of the tides, and to note their connection with the moon, and their periodic fluctuations.

Yet he was clearly no ivory tower academic. He was also a sailor and explorer, as evidenced by the undertaking for which he is most famed, which seemingly culminated in his becoming the first Mediterranean man to reach the fringes of the Arctic.

Pytheas most likely set sail from Massalia somewhere around 325 BCE. Exactly by what method he embarked is uncertain. It is possible he initially sailed by night in an inconspicuous vessel to evade the blockade that Carthage had deployed at the Pillars of Hercules, the entrance to and exit from the Mediterranean that we know as the Strait of Gibraltar. Or he may have set out across land, taking advantage of a shorter distance and well-established trade routes. Either way, we know he soon found himself in Britain.

Here he tarried awhile, documenting the inhabitants' mining and export of tin and visiting what he referred to as the island's "three corners" of Kantion, Belerion, and Orkas, which modern scholars have interpreted to be Kent and Cornwall in the southeast and southwest, respectively, and the Orkney Islands to the far north.

And then he kept going.

Exactly how far north Pytheas reached is unclear. No copies of the treatise he wrote following his journey, titled *On the Ocean*, survive. All we know is what later commentators wrote of his observations. Strabo, a geographer and writer in the first century BCE who was generally dismissive of Pytheas's claims about his journey, reported that the Massaliot sailed "six days north of

Britain" to a place named Thule, which lay "near the congealed sea." Scholars have debated over the ensuing centuries whether this was somewhere in Norway or possibly Iceland, but that it was at least on the fringes of the Arctic is given currency by the writings of the likes of Geminus, who in his *Introduction to the Phenomena*, published roughly one hundred years after Strabo, cited Pytheas as describing a place where "the night is extremely short: two hours in some, three in others, so that after the setting, even though only a short time has elapsed, the sun straightaway rises again."

We also have, via a writer named Polybius, a description from Pytheas of the ocean around Thule, the aforementioned "congealed sea" where "the earth and the sea and all things are together suspended ... existing in a form impassable by ship or foot."

Polybius cited this observation as a means of deriding a claim that he felt could not possibly be true; but today, we recognize that it is entirely plausible that Pytheas was the first European to describe sea ice.

———— ∾ ————

Part of the reason why the likes of Polybius and Strabo were so dismissive of Pytheas's account of his journey was likely that the details seemed utterly fantastic, so far removed from their everyday experiences in the Mediterranean as to not possibly be true, and that they ran counter not only to their own understanding of the world but also to the widely held contemporary perception of the farthest reaches of the globe.

For centuries, Greek mythology had held that the farthest reaches of the Arctic were inhabited by the Hyperboreans, who "lived beyond the North wind" and who, the poet Pindar wrote a century or so before Pytheas, held "feasts out of sheer joy. Illnesses cannot touch them, nor is death foreordained for this exalted race."

The truth, of course, is more prosaic; and in the centuries that have elapsed since Pytheas's trailblazing voyage, the history of European exploration of the Arctic has largely been one of optimism repeatedly rejected, of dreams perennially dashed, of lives lost in the pursuit of knowledge and glory. And while no shortage of expeditions have foundered on the metaphorical shores of the inland Arctic, crippled by the cold and confronted with starvation, the overwhelming narrative is of ships denied access, trapped, or destroyed by the sea ice that has perpetually acted as the region's sentinel.

Nor was it just explorers: for whalers and sealers, too, the ice of the Arctic was an enduring, menacing foe, perhaps never more than when it entombed and forced the abandonment of a fleet of thirty-three American whaling ships off northern Alaska in 1871. At the beginning of that decade, approximately two hundred American ships plowed the whaling grounds of the Pacific; by decade's end, more than one-third had been lost in the ice of Alaska and the Bering Strait, dealing a crippling blow to an industry that was already facing the challenges of declining whale numbers and growing alternatives to whale oil.

Unsurprisingly, sea ice became synonymous with its impenetrability, its powers of resistance, its capacity to not just thwart but destroy. For those who dared to test its resilience, it must at times have seemed not just an obstacle but an omen, a harbinger of impending doom.

It is an image reflected in literature and culture. For Samuel Taylor Coleridge's narrator in *The Rime of the Ancient Mariner*, its appearance engendered a sense of foreboding:

> *The ice was here, the ice was there,*
> *The ice was all around:*
> *It crack'd and growl'd, and roar'd and howl'd,*
> *Like noises in a swound!*

The ancient mariner, of course, was at the other end of the world, his Antarctic misadventure reportedly inspired by a real-life privateering voyage in which one of the officers had shot an albatross. Although most twentieth-century adaptations ignored it, the narrative framing of Mary Shelley's *Frankenstein* is set in an Arctic in which the region is mined for metaphor, for loneliness, unease, danger, and death. In the book's opening passages, its narrator, Captain Robert Walton, sets out for the North Pole with optimism and ambition: "I shall satiate my ardent curiosity with the sight of a part of the world never before visited, and may tread a land never before imprinted by the foot of man." He anticipates "a land surpassing in wonders and in beauty every region hitherto discovered on the habitable globe. . . . What may not be expected in a country of eternal light?" In time, however, his tone darkens, and predictably, ominously encroaching sea ice is the trigger: "We are still surrounded by mountains of ice, still in imminent danger of being crushed in their conflict. The cold is excessive, and many of my unfortunate comrades have already found a grave amidst this scene of desolation." Victor Frankenstein, who Walton discovers pursuing the creature he has created across the Arctic's icy expanse, dies onboard, prompting the creature to tell

Walton that it intends to build a funeral pyre for itself in the Arctic's farthest reaches, ensuring its burned corpse "may afford no light to any curious and unhallowed wretch, who would create such another as I have been."

A place of ice and death, where monsters commit fiery suicide: readers of Shelley's novel, who were and remain legion, could scarcely be blamed for coming away with an image of the Arctic as little more than a collector of souls, and the omnipresent sea ice the cudgel with which it bludgeons its victims. The book was published in 1818, just as John Barrow received permission and funding to launch a new wave of Arctic exploration for the British Admiralty. A little over a decade later, John Ross's exploration of the Northwest Passage onboard the *Victory* turned into a four-year ordeal as he struggled vainly to escape the ice's clutches; within thirty years, the *Erebus* and *Terror* disappeared, leaving no survivors and yielding tales of cannibalism. Small wonder, then, that the notion became embedded in the collective psyche that the Arctic was a hostile place and its ice a foe best avoided and certainly not to be celebrated.

When the purpose of traveling through the Arctic is to get from point A to point B, when the Arctic is nothing more than a blank space on the map begging to be colored in, its very nature becomes a hindrance and an inconvenience, and the defining element of its waters becomes a foe to be feared. Even Fridtjof Nansen, among the more cerebral and scientifically minded of explorers, was moved to exclaim: "Our hearts fail us when we see the ice lying before us like an impenetrable maze.... There are moments when it seems impossible that any creature not possessed of wings can get farther, and one longingly follows the flight of a passing gull, and thinks how far away one would soon be could one borrow its wings."

It is not an enormous leap from the accounts of explorers wishing the sea ice that surrounded them would simply disappear to a world in which there are plenty who welcome the fact that sea ice is showing the first signs of doing exactly that. When it has been nothing but an obstacle, when its purpose seems solely to thwart and destroy, when it is a source of terror rather than wonder, its gradual disappearance and the emergence in its place of the network of shipping lanes that have been sought for centuries is a cause for satisfaction rather than regret. At worst, it is a shrug of the shoulders at the twists and turns of fate.

To be fair, anyone who has spent much time in polar seas will recognize numerous sentiments with which to emphasize in Nansen's despairing words. In both the Arctic and Antarctic, I have experienced the terrifying power of sea ice. In the Arctic, I have been onboard an icebreaker that was uprooted from anchor by an enormous ice floe driven by powerful Beaufort Sea currents; in the Antarctic, I have experienced the singular fear of being hemmed in on all sides by pack ice as far as the eye could see and wondering how we could possibly escape to freedom. Sea ice is truly an awesome, powerful, and not infrequently terrifying phenomenon. It is also wondrous—as Nansen also was moved to declare, "a shifting pageant of loveliness."

It is, additionally, the basis of life in Arctic waters. Without melting sea ice releasing the diatoms that have embedded themselves in it, there is no explosion of marine life in the spring. And for Inuit across the Arctic, the fundamental element that has sustained their way of life and that of their ancestors for millennia is gone, an entire people denied the element that protects their coastal communities from storm surges, that provides a vital form of intercommunity transport, that not only acts as their hunting grounds but also sustains the wildlife they set out to hunt. It is a slowly unfolding cultural genocide conducted from afar because people in the south did not want to acknowledge the impacts of unfettered fossil fuel use or did not want to surrender the comforts that came with it, objected to the notion of implementing reform and regulation because it clashed with their political philosophy, rejected the very concept of human-caused climate change because it offended their religious sensibilities, or simply didn't care enough because they prioritized present-day personal riches over future global conflagration, or looked at a map projecting future shipping routes through the Northwest and Northeast Passages or across the North Pole and decided they liked what they saw. Why worry about the extirpation of polar bears or seals or the disappearance of a culture when packages can reach Europe from Asia a couple of days more quickly?

It would be wrong to suggest that the Inuit view on Arctic shipping is monolithic; for many communities, life is hard, and the effects of isolation are real. The limited ship traffic that presently does exist is for many the primary or even only means of contact with the outside world, and the prospect of greater amounts of it—and with it the potential for much-needed income— is undoubtedly attractive. Equally, Arctic peoples—like those anywhere— are as entitled as any to share in the monetary, medical, and material benefits of progress. But there is a sordid history of northern peoples being denied the life they are accustomed to, the one with which they are familiar and culturally attuned, by those who sought to "civilize" them or who thought they knew what was in their best interests. Requiring them to accept and adapt to a world without sea ice may be more insidious than relocating them against their wishes, refusing to allow them to speak Inuktitut, or forcing them to attend Christian schools, but its potential impact is no less profound.

It would be wrong, too, to suggest that people of the Arctic view sea ice as a solely benevolent and beneficial entity or to retreat into the canard of well-meaning Westerners that Inuit and other Indigenous peoples live in some kind of symbiotic harmony with nature. European explorers were correct in their assessment that sea ice has the capacity to be a powerful and destructive force; they discovered as much at considerable personal cost, and few people can be more aware of its perfidious and dangerous nature than those whose lives are spent on and around it (even if the greatest danger from sea ice for Inuit often comes less from the ice itself than from the cracks that may be hidden within it or the fog that can descend while they traverse it).

Even so, even when acknowledging the dangers of a life spent on and revolving around sea ice, Inuit frequently speak of it with an element of reverence engendered by gratitude rather than the outright, if understandable, fear expressed by interlopers. The dichotomy is expressed well in a passage from *The Meaning of Ice*, a comprehensive volume edited by Shari Fox Gearheard and colleagues that showcases first-person narratives on sea ice from a trio of Arctic communities:

> Sea ice is the bridge connecting people and the path to what is next. Sea ice travel takes us to places on both land and sea. The ancient trails we retrace, and even the new trails we create, are places themselves. Traveling sea ice is a part of our individual and family stories and histories, which we continue to tell and add to. Along the way we must always respect the potential hazards of being on sea ice. . . . Freedom is a gift, but one to be treated with great respect.

Added Joelie Sanguya, a hunter and *qimuksiqti*, or dog-teamer, from Nunavut in the same volume:

> We work, travel, hunt, fish, sleep, live, play, have games, and picnic on the sea ice, which is a very important part of our

culture. As children, we would . . . mimic being a seal on the ice, or a polar bear, or caribou. These games were how we used to learn to use and take advantage of the existence of sea ice and how it provides us with food and a place to travel. Through these games we would get to know ourselves better.

There is a tremendous amount of flatness, of open space provided by sea ice when it forms. You can't find anything else like it in the world except where sea ice freezes. Some people cry not for sadness but for *joy* of living with this life-giving wonder. It provides for us and heals our hearts and minds as we travel upon it.

In 2005, Sheila Watt-Cloutier, then the head of the Inuit Circumpolar Council, launched the world's first international legal action on climate change: a petition to the Inter-American Commission on Human Rights alleging that unchecked emissions of greenhouse gases from the United States violated Inuit cultural and environmental human rights as guaranteed by the 1948 American Declaration of the Rights and Duties of Man. Specifically, it violated one right above all, a right that most people in the lands south of the Arctic, and indeed many of those who have ventured into polar realms only to leave in short order with their tails tucked between their legs, might struggle to comprehend: the right to be cold.

In a 2018 panel organized by the Clean Arctic Alliance, subsequently posted online as "A Message from the Arctic," Watt-Cloutier outlined the importance of sea ice and its preservation to Inuit specifically and the Arctic more generally, and beyond that the importance to the world at large of looking at the region as something more than a future network of passages for commercial shipping:

> The ice is our life force, the ice is our university, the ice is our supermarket for highly nutritious food, the ice is about the health and well being of an entire people who live at the top of world who have already gone through many historical traumas and who now face the possibility of even more vulnerable situations. It is not only about polar bears, it is about families and communities.
>
> These challenges we face are not of our doing. We have benefited the least from industry and yet we are one of the

most disproportionately negatively impacted by the effects of globalization. How ironic and sad that a people who have lived sustainably for millennia and far from the source of pollutants and greenhouse gases would bear the brunt of their damaging effects.

For too long we have asked the world to stop hurting our way of life, and for too long the world has responded that it is too expensive to stop bringing harm to our way of life. Inuit and other vulnerable peoples are being asked to pay the price for the unsustainable choices most of the world continues to want to maintain. Inuit and others are becoming the collateral damage as a result of irresponsible political actions ... or inactions.

Everyone benefits from a frozen Arctic and that everything is connected and we can no longer separate the importance and value of the Arctic from the sustainable growth of economies around the world.

Everything is connected through our common atmosphere, not to mention our common spirit and humanity. What affects another, affects us all. We know that as the Arctic melts, other places such as the Small Island Developing States are sinking.

In the international arenas, where I have personally been involved, the language of economics and technology is always calling for more delays. We are constantly reminded how taking action on greenhouse gas emissions will negatively impact our economy. I understand this same lame excuse, which is a very outdated card to play at this stage with our climate crisis. ...

Instead we must now be thinking about and reframing all these issues in terms of fundamental human rights. This approach of focusing climate change solely to economics tends to "silo" or separate the issues from one another, as opposed to recognizing the connections between rights, environment, health, economies and society. ...

It's a human issue, a human rights issue, and the wisdom and solutions lie within us. Together we can do this.

CHAPTER 14

Journey to the Top of the Earth

The North Pole is a destination without a marker.

There are no mountains, no permanent topographic features of any kind, just a seemingly endless expanse of jumbled ice and intermittent patches of open water stretching as far as the eye can see. To be at the North Pole is to feel as removed from the rest of humanity as it is possible to feel, isolated in a harsh environment thousands of miles from civilization and warmth.

Except when I was there.

When I was at the North Pole, there was a replica British-style phone booth, a barbecue, a line of people taking turns to leap into a patch of freezing water, and another group of people playing soccer. Even Santa was holding court, although on this occasion Santa sounded—and, beneath the beard, looked—suspiciously like a German woman named Roswitha.

And there was a ship. A very large ship, in fact: a five-hundred-foot, 28,000-ton, red-and-black nuclear-powered Russian icebreaker called *50 Let Pobedy*—or, in English, *50 Years of Victory*. This was the ship that had transported us all—me, Roswitha, the polar skinny-dippers, the soccer players, ten dozen or so tourists from around the world, journalists, scientists, and, for good measure, the Russian park rangers who guarded our perimeter in case the barbecue attracted any curious polar bears.

The Soviet Union had first dispatched an icebreaker to the North Pole in 1977. The *Arktika*, whose journey was launched to mark the sixtieth anniversary of the start of the Russian Revolution, was the first surface vessel to reach the Pole; in the decades since, several vessels had completed multiple journeys to the northernmost point on Earth as the ice had started to thin and more people with more disposable income saw it as a worthy vacation goal. Our voyage was the 123rd.

For decades, men and women had striven to be where I stood now and had struggled on skis, hauled sleds, and endured a litany of miseries in pursuit of that goal. In contrast, all I had to do was board a flight to Helsinki, Finland; join my fellow passengers on a charter to Murmansk, Russia; and from there kick back and enjoy the scenery, the wildlife, and three multicourse meals per day while the ship's crew did all the work of getting us to our destination. The most meaningful decision I had to make was whether to have red or white wine with dinner, and the greatest struggle was climbing out of my bunk early in the morning for yet another call from the bridge to see yet another polar bear—or bowhead whale, or walrus.

But it didn't matter. At that very moment, everyone else on Earth was to the south of us, beyond the horizon that bounded our icy surroundings. The ease of the journey did not diminish the significance of its destination; as I drank in the scene, I reflected on how remote we were and on the fact that even now, in an age when it is possible to achieve by hitching a ride on a giant ship what once was the sole province of those prepared to commit their life to the cause, it was a mighty small club indeed of which I was a now a member.

As of that moment, I was one of the very few people in history who could say that they had stood on top of the world.

If, as Robert Peary claimed, he and Matthew Henson were the first men to stand at the North Pole on April 6, 1909, they did so after more than a month of hard slogging from Canada's Ellesmere Island. Sixty years later, when Wally Herbert became the first person universally acknowledged to have walked to the Pole, he and his team had been on the ice for fully a year, having been forced to make camp over the long Arctic winter and wait until currents carried the sea ice in a favorable direction.

In 2017, those of us onboard *50 Let Pobedy* made the trip in under a week. We arrived in Murmansk on the afternoon of August 1 and began our smooth, steady journey northward that evening. We had been instructed not to take any photographs until we were clear of the port, which is home to the Russian nuclear fleet; but as the ship slipped from its moorings, the mood lightened, and the celebratory on-deck toasts loosened inhibitions, that admonition was soon forgotten. Cameras and cell phones clicked away as *Victory* (to use the affectionate and time-saving name by which our ship was referred to by those onboard) eased quietly out of harbor, into Kola Bay, and northward into the Barents Sea.

Within two days, we had reached Franz Josef Land, an archipelago of 192 islands that is the most northerly land in Eurasia—at its northernmost point a mere 560 miles from the Pole. We would visit it again on the way south; but in between, once the archipelago had slipped over the horizon astern of us, we would see no land. The journey to the Pole was devoid of craggy cliffs and stunning vistas, the only variants the extent and thickness of the ice floes that surrounded us, the amount of water that separated them, and the wildlife that crossed our path or tailed in our wake. Our first sightings of ice came as we made our way past the archipelago, but it was on the evening we left the islands behind us that the sea ice shifted from being an occasional interloper to the dominant feature of our surroundings. The ship rattled and shuddered as it entered the Arctic Ocean ice pack, crushing and plowing through the floes that had the temerity to stand in its path. Smaller ones were tossed casually to one side, but even the larger sheets offered little to no resistance. I leaned over the bow and watched as a crack would appear in the ice and race ahead in jagged fashion, splitting a floe asunder and then widening and ultimately separating the floe into two or more pieces as *Victory* waltzed arrogantly through.

The following morning, we saw polar bears. The first sighting came sometime before 6:00 a.m., and the next followed an hour or so later, when the initial male was joined by a second, picking his way among the ice floes, sniffing the air, and investigating the unfamiliar aromas that were gently wafting their way toward him. If either bear was in any way concerned by the presence of the nuclear-powered icebreaker, neither registered any visible displeasure.

The initial bear, rotund and healthy, spent most of his time doing what polar bears often do: lying at the edge of a floe, conserving his energy, and perhaps hoping that a seal might pop out of the open water. The second headed closer, but then, as if belatedly becoming aware of the presence of the first, displayed caution and kept his distance. We watched them for a while until those onboard had had their fill, and then the ship ghosted slowly and silently away.

They would not be the only bears we encountered during our voyage. On our way to the Pole and again heading south, we saw in total close to a dozen, many of them in and around the Franz Josef Land archipelago, where currents and narrow channels pushed ice floes together and where the islands gave the bears the opportunity for rest ashore. One was in repose on a thick jumble of shore-fast sea ice along the coast of one of the islands. Another, also in Franz Josef Land, was resting on a bluff just below a cliff face full of kittiwakes, having perhaps gorged on unsuspecting birds and their chicks, its dozy presence enough to prevent us from going ashore to check out the scenery. But the majority were in their favored domain, wandering across expanses of fractured sea ice, including a mother with her young cub striding confidently in her wake. Fractured sea ice is ideal polar bear habitat, providing solid ground over which to roam and open water through which a seal head might pop at any time. Indeed, while we watched the mother and cub, the adult bear at one point darted for an open lead in the ice, presumably attracted by the sight, smell, or sound of potential prey. Those assembled on the deck were denied the chance to see her make a kill, however; it was evidently a false alarm, and mother and child continued on their way, wandering across the expanse of ice into the distance.

Victory makes only five trips to the North Pole with paying passengers each year, chartered alternately by Quark Expeditions and Poseidon Expeditions; it spends the bulk of its life breaking through the ice of the Northern Sea Route, opening pathways at the head of convoys of cargo and container ships. It is a working ship, not a cruise liner, and the accommodations—notwithstanding the journey's starting price tag of $27,000 for the eleven-day round trip—reflect as much. Still, if the cabins weren't ornate, they were functional—mine, which I shared with my friend and colleague Geoff York from Polar Bears International (PBI), was barely large enough to swing a small Arctic mammal, but it had a pair of bunks, a desk, a small bathroom with shower, and plenty of storage space—and was replete with nice touches, including daily housekeeping service topped with a nightly treat of *50 Let Pobedy*–branded chocolate. And while the ship, with a blocky red superstructure atop a black hull, carried the air at first blush of a floating communist-era apartment block, it was plenty comfortable. There was a library, a bracing below-decks seawater swimming pool, a sauna, a basketball court, and a small gym. A gift shop, run by the aforementioned Roswitha, sold a variety of books, clothing, and other souvenirs. And a bar with a panoramic window allowed all onboard to watch in comfort as the ice ahead was pummeled by *Victory*'s strengthened bow. Each evening, we all convened for briefings on the day's sights and sounds and the following day's schedule in a large theater-cum-lecture room toward the stern, which was also the site of multiple presentations throughout the day—from Geoff and PBI executive director Krista Wright on polar bears and on climate change, and from members of Quark's onboard team of experts on multiple subjects ranging from icebreakers to seabirds to Austrian–Hungarian explorations of Franz Josef Land.

When I first traveled to the Arctic twenty years before, I shared a cabin with the cook, chipped in with chores, stood some watches on the bridge, and shared the best part of three months with perhaps thirty sailors, journalists, and scientists. This time around, I rarely saw much of the Russian crew; our primary interactions were with the dozen or so Quark staff, a multinational and multilingual cast each with a specific specialty—historian, ornithologist, helicopter pilot—but all of whom also functioned as general tour guides, boat-drivers, answer-givers, problem-solvers, and hand-holders, tasks they all discharged with impressive efficiency and hospitality. Then there was the positively polyglot passenger manifest: about one-third were Chinese, and there were small groups of French, Japanese, and Austrian passengers, at least one New Zealander and one South African, a pair of British freelance journalists, a couple of French-Canadian filmmakers, and a handful of Americans, including Krista, Geoff, and myself. There were families, groups of friends, and solo travelers, most middle-aged or older. Several were retired, others working but wealthy, some who had splurged savings and inheritances on the trip of a lifetime, and a handful of inveterate travelers looking to add another exotic notch to their belts.

The other big difference was the ship itself. In 1998, I had been on a thousand-ton, ice-strengthened converted sealing ship; it coped admirably with all but the very thickest of ice floes, but its captain approached them cautiously, slowing down on approach, avoiding them whenever possible, pushing them out the way when he could, and riding on top of them and breaking them when necessary. Onboard *Victory*, Captain Dmitry Lobusov—who, with his gray beard and stiff, serious bearing looked every inch the captain of a Russian nuclear vessel—had no such cares; even as the dinner plates and wine glasses rattled, the ship plowed through the ice as if it were wet tissue paper.

"It's kind of remarkable, really, to be on a ship that just slices through ice with barely any trouble at all," noted Geoff one evening as we looked through the window in the bar at a frozen seascape of ice being beaten into submission.

To a large extent, that was a consequence of *Victory*'s immense strength, of engines that generate an almost unfathomable 75,000 horsepower, and of a stem of sixteen-inch-thick steel. The ship was built to blast through sea ice. But it was not the only factor.

It was not just that the sea ice was interspersed with frequent expanses of open water. August is high summer in the Arctic, the time when its sea ice coverage is rapidly progressing toward its annual minimum, so it would be expected that there would be far less of it than in, say, March. Patches of water amid the ice at this time of year were scarcely without precedent, even at or close to the North Pole itself. Even so, for those who had made the trip on multiple occasions over the years, the difference was notable.

"The change is really striking," said Colin Souness, a Scottish glaciologist who works onboard as a Quark guide, scientist, lecturer, boat-driver, and human Swiss Army knife. "And it really drives home how quickly the ice responds to even slight changes in temperature. At one degree below freezing, ice is resilient and quite rigid. But as soon as it gets to zero [Celsius], it becomes so much softer, and you can see that with the ease with which the ship was just gliding through it."

Captain Lobusov, while dubious that the cause was anything other than natural cycles, acknowledged the changes. "I have been working in the area for thirty years, and been doing North Pole voyages for twenty-four years, and I've seen many changes in the ice conditions," he said during a sit-down in his office with the small group of journalists on board. "Now, we hardly see the thick, multiyear ice we used to have two decades ago; as we approach the North Pole, you can see we have many stretches of open water."

The thickness of the ice—or lack thereof—that Captain Lobusov mentioned was especially striking, especially disconcerting, and especially notable. According to the Polar Science Center at the University of Washington, 2017 had the lowest annually averaged sea ice volume on record. We didn't know that at the time of our cruise, of course: we just kept looking out at the ice, allowing ourselves to grow briefly excited by hints of it becoming thicker and to wonder whether, just over the horizon, we would start to see gnarled, jumbled masses of ice that had grown over year upon year.

We watched, and we waited, and we searched, but they never came.

———— ∽ ————

We reached the North Pole at a little after midnight on the night of August 5 to several long, loud blasts of the ship's whistle.

As the destination grew ever closer, Solan Jensen, a fiercely organized Alaskan with a Zen mien who served as Quark's assistant expedition leader, counted down the remaining time and distance over the ship's public address system like a NASA controller. Some chose to spend the approach in the wheelhouse, taking selfies with the GPS as it ticked ever closer to 90°N. The bulk of those onboard assembled at the bow, where music was playing and vodka and champagne were pouring. After starting my evening in the former, looking down on the massed ranks of ready-to-partyers, I soon switched locations.

Our arrival at the Pole was met by an eruption of joy from the 130 or so passengers and Quark staff. There were hugs and cheers, selfies and ussies, toasts and more toasts, dancing and laughing and tears of joy, and as I stood there, the tsunami of delight washed over me, too, as I connected with the emotional resonance of where we were.

Suddenly, in that moment, if only for a moment, it didn't matter that I had been merely an idle passenger while the ship's crew and Quark's staff had made everything possible. I could even temporarily overlook the fact that, as I peered over the side of the ship, the patches of open water were as notable as the expanses of ice. I was at the top of the world, in the very heart of the Arctic, just about as far away from any population center as it was possible to be in the Northern Hemisphere, a destination that had been shrouded for centuries in legend and mystery and myth.

It was hard not to think at that moment of all those who had struggled and failed to force ships through the ice in a vain attempt to reach this very spot, who had died of frostbite or hunger or scurvy while searching for the Northwest Passage or the Open Polar Sea, who had pined for home even as the realization slowly dawned that they would never again see their loved ones but would join the roll call of victims of the Arctic's brutal hostility.

Had any of those pioneers had access to a vessel remotely as powerful as *50 Let Pobedy*, of course, the history of Arctic exploration would have been far more a catalog of triumph and far less a chronicle of glorious failure. Whatever the advisability of Russian-built nuclear power plants bludgeoning their way through Arctic ice, *Victory*'s capability and endurance were undeniable, to the considerable and understandable pride of its chief engineer, Vladimir Yudin.

"With five hundred kilograms of uranium, we have enough fuel for five to six years," he told a few of us during a selective tour of the engine room— selective in that it pointedly and purposefully avoided the nuclear control room. "Two desalination plants each produce one hundred tons of fresh water per day, more than is consumed even when the standard crew is joined by one hundred or more paying passengers. Our only limitation is the tiredness of the crew and the capacity of food storage."

For some, that left a certain emptiness to the achievement. For all that Lobusov had offered us all as we stood out on the ice, "Congratulations on achieving your dream," the absence of peril or indeed any kind of sacrifice or jeopardy on the passengers' part prompted one, Californian Palle Weber, to offer, sotto voce, "I'll be honest; it feels a little like we summited Everest by helicopter."

For others, however, the comfort of the conveyance was an irrelevance. Simply to have reached this point, a location that has been and remains for the great majority of humanity a place of inaccessibility, was an achievement and a cause of reflection in and of itself. Ranjan Sharma, a well-traveled allergist from North Carolina, posed for pictures and then confessed, "I had to break away for a bit. I needed some time alone, to reflect on where we were and what it meant and all the things that had had to happen in my life for me to be here."

The marvel and majesty of the moment, and indeed of the whole journey, was best summed up by Jensen, the assistant expedition leader, as we sat in the lounge one afternoon on the voyage north.

"The experience of hearing and watching the ship break ice is as mesmerizing as watching fire," he shared, eyes closed to better enable him to contact and convey the appropriate emotions. "Its sounds defy language in many ways, because it's cracking and crumbling and heaving and bursting; and the visuals of cracks and leads opening up in the ice, and where and how that happens, and the speed at which it happens, is something that everyone is taken away with—at least in the moment, if not for days and days at a time. And then there's the experience of the ship parking in the ice in the middle of the Arctic Ocean at 90°N. We lower the gangway and everyone climbs down onto a meter and a half of ice on top of four thousand, five thousand meters of water. Having the opportunity to reflect on where you are, looking out and knowing that over the horizon—in whichever direction you look—are all the people on the planet, all the continents and islands, all the cities and cars. . . . I think it's quite hopeful. I think it delivers people a sense not of isolation, but community. I think it's very powerful."

Lobusov was forced to spend several hours overnight steaming around the North Pole in search of an area of ice thick enough for him to park the ship. Once he had done so, and while the rest of us slept, the Quark team set up the site to be as safe as possible for passengers to disembark. Flags marked a trail deemed sufficiently secure, across thick enough ice and away from treacherous water. For the more daring sorts, an open patch of water next to the ship provided an opportunity to leap—while securely tethered—into the freezing Arctic Ocean and emerge as swiftly as possible. And that replica British phone booth marked the spot at which Quark staffers stood with a satellite phone so that every one of us could call a loved one—briefly—from the North Pole.

I wandered off for a moment of solitary contemplation. It wasn't especially cold—about 30°F; it was, after all, the height of summer—but a sharp wind, uninterrupted by any geology, drove across miles of sea ice, and I buried myself in my parka as I surveyed the scene.

We had been parked at—or, more accurately given the conditions, as close as possible to—the Pole for twelve hours or so; we would remain for only a couple hours more until we began the journey south. We had eaten, we had drunk, some of us had dunked, we had strolled on the ice, we had taken plenty of photographs. Tasks fulfilled, most of the passengers had now retreated to the comfort of *Victory* or were preparing to; only the hardiest, youngest soccer players remained, looking as if they could stay for much longer yet.

Soon all signs of our presence would be gone: the barbecue tables, the soccer goal, the phone booth, and then even *50 Years of Victory* itself, disappearing over the horizon. In two weeks, it would return for the last time in the season, and then it would be gone again, taking with it the last signs of life at this most remote of locations. Summer would be over, the temperature would plummet, and the North Pole would surrender anew to the cold, the dark, and the quiet.

Epilogue: Futures

In the English village of Greenhithe, on the banks of the River Thames estuary, there is a pub.

The exact year in which it was built is uncertain, although it was certainly up and running in 1661, when two murderers sought refuge there but were arrested and later hanged for robbing and killing a man named Cossuma Albertus, an impoverished Polish nobleman who fashioned himself as a prince of Transylvania.

The pub was known at the time as the King's Head; but, in 1742, possibly out of belated deference to the decapitated Charles I, it was renamed the White Hart. That was the appellation it carried until the early years of the twenty-first century, when new landlords changed its name to honor the man who historians, professional and amateur, would discuss at the bar and who was surely its most celebrated former patron.

It is now the Sir John Franklin.

It was in one of the boarding rooms upstairs that Franklin reportedly lay his head on dry land for the last time, on May 18, 1845, before rowing out the next morning to join HMS *Erebus* as it and *Terror* set sail for the Northwest Passage. The scene then would have been far different from the one that presented itself as I stood on the bar's patio on a rainy morning in January 2024. Today, an occasional bulk carrier steams past the industrial backdrop on the other bank of the river as, to the west, cars and trucks stream endlessly in and out of London via the Dartford Crossing; but two centuries before, it would have been thronged with vessels carrying passengers and goods or even preparing to set sail on a voyage of Arctic discovery.

Outside of the pub's name and its nautically themed decor, there was, for 179 years, no obvious indication that this was the spot from which *Erebus* and *Terror* had departed. The public causeway that once ran along its side and the draw dock at which a multitude of boats had, over time, been moored have long gone, and the riverside has been transformed by a seawall and other flood defenses. It is difficult to picture the scene as it would have appeared in 1845 or to imagine this patch of river as the launching pad for one of history's most infamous expeditions.

But on this January morning, Franklin and his men were about to receive official recognition of their connection to the area. Watched over by invited guests, naval cadets, local councillors, and the mayor, and protected by a black umbrella, Sir Michael Palin—once of the Monty Python comedy troupe and in his later years enjoying a second career as a respected travel writer and broadcaster—unveiled a plaque commemorating their journey.

For Palin, who had written a book about the *Erebus* following the discovery of its wreck, it was a moving experience.

"People said farewell to Sir John Franklin and his sailors here," he mused. "They were going off into the unknown and they never came back. This was their last port of call and the last time a lot of their families would have seen them. It is freighted with significance, as it was the last time they would celebrate together. It's very emotional."

Much about the setting would have been utterly alien to Franklin and his crew: the airliner that had carried me across the Atlantic, the motorized vehicles parked in front of the pub, the televisions on the pub wall and the Premier League soccer they broadcast, the global pandemic that had delayed the plaque's unveiling by several years, the notion of being commemorated by a man most famous for an extended riff on a dead parrot.

Would he have been similarly baffled by the fact that almost two centuries after his expedition, the Northwest Passage remained traversed by only a relatively small number of vessels, or that its status had been a matter of almost continuous contention for much of the time since he had departed Greenhithe? And what would he possibly have made of the reason for growing enthusiasm about the Passage's viability, or that of the Northeast Passage—also unnavigated during Franklin's time—or even a pathway across the North Pole, which he himself had attempted and failed to find? Would he have struggled to come to terms with the fact that not only was the planet's very climate undergoing a transformative change, and not only was that change being brought about by humanity itself, but also that the overall response to the slowly unfolding planetwide disaster was essentially a global shrug? That even the efforts being made to reduce the pollution that was altering the very fabric of life on Earth were largely recalcitrant and being conducted at a pace and level of urgency entirely discordant with the scale of the unfolding disaster? Perhaps he might note an uncomfortable echo of the misadventure that took his own life and the lives of 128 others: multiple warnings and evidence from previous misadventures ignored and met with a triumphalist hubris until a slowly unfolding disaster came to fruition.

———————∞———————

Franklin's era of Arctic exploration was fueled in part by competition between nations, by Britain's desire to achieve priority and earn both glory and commercial advantages at the expense of their great power rivals. That rivalry is echoed in contemporary approaches to the passages, although it does not necessarily follow that it has reached or is in danger of reaching the more extreme levels postulated in much popular discourse.

For example, a quick Google search for "Arctic cold war" generates a plethora of headlines ranging from the somber and understated to the borderline breathless: "The U.S. must wake up to rivalry in the Arctic." "The winds of the new cold war are howling in the Arctic." "In the Russian Arctic, the first stirrings of a very cold war." "Melting Arctic could lead to U.S.-Russia war." "China's expanding Arctic ambitions challenge the U.S. and NATO."

The parameters of the search inherently favor more sensational results, but even allowing for the fact that headline writers and article writers are not one and the same and that the latter (ideally) favor more nuance, the tone of much such coverage is similar: as Arctic ice melts and the region becomes more accessible, the region's powers jockey for territory and influence—which is all perfectly reasonable—and that battle may manifest in military confrontation, which is altogether more conjectural.

It's an appealing narrative, not least given that it fits so squarely into the general perception that countries, even nominal allies, act in perpetual competition: for influence, cultural or linguistic hegemony, trade advantages, military might, sporting prowess, productivity and wealth, military might, natural resources, territory . . . the list goes on. Factor in the identities of the key players in this perceived conflict—the United States and Russia, with a dash of China—and the storyline becomes almost irresistible.

As we have seen, tensions have existed over Arctic sovereignty and rights-of-way, and they still do. It seems reasonable to assume that as ice continues its retreat, such tensions are likely to ratchet up. So far, in the grand scheme of things, it's all been rather mild: a strategically dropped flag here, a territorial claim there. But these are the kinds of disagreements—over territory and borders, over seaways and access to minerals, oil, and gas—that can fester unresolved, smoldering in the undergrowth until a sudden change in conditions causes them to catch fire.

Then again, the response of those competing Arctic nations to outside calls for the imposition of some form of new governance regime on the region was to circle the wagons and strengthen cooperation: not exactly indicative of a group of countries in irredeemable conflict over Arctic resources.

That said, the geopolitical wheel can spin quickly, and it would be wrong to ignore a growing military presence in one part of the Arctic and calls for a response elsewhere.

"Russia now has six bases, 14 airfields, 16 deep-water ports, and 14 ice-breakers built," Admiral Daryl Caudle, commander of the US Fleet Forces Command, said at a 2023 seminar sponsored by the Polar Institute of the Woodrow Wilson International Center for Scholars and the Center for Maritime Strategy, as cited in an article for *Seapower* by Richard Burgess. "They dominate the Arctic geography and possess the corresponding ability to dominate in capability and infrastructure.... They have an active defense system that has high readiness, mobility, and firepower in the Northern Fleet.... They have long-range, precision-guided strike weapons especially focused in and near the Kola Peninsula."

Media accounts have picked up on such statements and conflated them with melting sea ice, with the Northwest and Northeast Passages, and with the alleged wealth of unretrieved riches that lie in the Arctic Ocean's soon-to-be-exposed seabed. But the connection is not as clear-cut as the casual associations might imply.

A few paragraphs after noting that "as a shrinking ice cap opens up new sea lanes and resources, the Arctic is becoming strategically more important," a November 2022 Reuters report by Jacob Gronholt-Pedersen and Gwladys Fouche observed that "the shortest path by air to North America for Russian missiles or bombers would be over the North Pole." Both statements are objectively true, but the two are unconnected. In December 2023, Holly Williams and Analisa Novak of CBS reported:

> The melting polar ice caps have opened new shipping routes and exposed untapped reserves of oil and natural gas. Russia is testing hypersonic missiles, capable of evading American defenses, in the Arctic. This August, a joint Russian and Chinese military flotilla was observed patrolling waters near Alaska.

Again, all true enough. But one does not necessarily lead to the other. Russia's militarization of its Arctic region, although very real, is most likely part and parcel of Moscow's ongoing industrialization of its northern realms combined with the same paranoia over others' intentions that prompted Tsar Mikhail Fedorovich to close the Mangazeya Sea Route several centuries ago. Indeed, Caudle himself acknowledged in his presentation that Russians "do have legitimate sovereign interests" in the region. While Russia's neighbors, including in the Arctic, understandably view any military growth by the bear next door with trepidation—with Finland and non-Arctic Sweden even abandoning decades of neutrality to join the North Atlantic Treaty Organization—that concern is more a consequence of Russia's behavior in theaters farther south, specifically its aggression toward Ukraine. Tensions may well be growing in the region, but while talk is rife of potential future conflict *in* the Arctic, it does not necessarily follow that it would be *of* the Arctic. The rapidity with which missiles can fly across the North Pole between Russia and North America is completely unaffected by whether the ocean beneath them is open water or covered with heavy pack ice. It is worth emphasizing: there is, to all intents and purposes, presently virtually no traffic through the Northwest Passage. There is essentially no Transpolar Route at all for now. The Northern Sea Route, largely for reasons of Russia's own making, is presently primarily a passage for domestic and destination shipping. The undersea resources that are largely believed to be awaiting retrieval as retreating ice permits greater access are largely found in Arctic rim countries' territorial waters anyway.

The ingredients for regional conflict are simply presently lacking, frenzied speculation notwithstanding. It is hard to escape the conclusion that briefings to the contrary are another example of countries—and in this case, their military and foreign policy establishments—finding their comfort zone: linking a potential future challenge (in this case, a melting Arctic) with a possible security threat, accusing others of acting in bad faith or aggressively, and insisting on the right to respond in kind. Add to that a media that is generally much more comfortable with simple narratives than nuanced analysis and the ingredients are in place to be whipped into a perfect storm of frenzy.

It is important not to be overly Pollyannaish: as we have seen, tempers have risen over the prospect of Arctic sea lanes, and it is not difficult to envisage a future situation that could spiral rapidly out of control: an American administration uninterested in long-standing global norms sending a warship through the Northern Sea Route for no other reason than to establish its muscular bona fides; a late-stage Putin, increasingly untethered from reality, seeking to bolster popularity domestically by overreacting and seizing it, almost certainly putting Moscow on the wrong side of international law and prompting condemnation from the bulk of the United Nations; an emboldened Washington doubling down; Russia's Arctic military forces being placed on full alert . . . none of it is implausible, particularly when relations between the two nations are at a post-1989 low. The invasion of Ukraine has led to Russia's exclusion from the Arctic Council as well as other regional forums in the Arctic; while the rationale for such action is understandable, the lack of open cooperation and sharing of information is more likely to lead to suspicion, uncertainty, and a readiness on all sides to assume the worst. There are precedents of sorts: ultimately, Gavrilo Princip shooting Archduke Franz Ferdinand was all it took to spark a conflagration between two sides itching to settle differences that, more than a century later, make little or no sense and that sent twenty million people to their deaths over four years. The flashpoint need not be the Northern Sea Route or a territorial claim in the Arctic Ocean but could be somewhere else entirely, perhaps on the archipelago of Svalbard. Although it is Norwegian territory, the archipelago has an usual political structure and a strong Russian presence stemming from decades of a considerable Soviet mining presence, with the town of Barentsburg, for example, still containing around 450 Russian citizens. In 2023, some of those residents staged a May 9 "victory day" parade through town, and the following year, at least three Soviet flags were spotted flying above Barentsburg and the abandoned Soviet coal mining town of Pyramiden. The latter act in particular brings to mind Argentine scrap dealers raising their flag in South Georgia in 1982, setting in motion the bloody and senseless Falklands/Malvinas war between Argentina and the United Kingdom.

Unfortunately, once momentum toward the prospect of conflict develops, it can become self-perpetuating. And so, the US Army released a document in 2021 titled "Regaining Arctic Dominance," and Mike Baker of the *New York Times* noted in March 2022 that "while the United States has denounced Russia's aggressive military expansion in the Arctic, the Pentagon has its own plans to increase its presence and capabilities. . . . The Air Force has transferred dozens of F-35 fighters to Alaska, announcing that the state will host 'more advanced fighters than any other location in the world.'" With each document, each position paper, each article, each piece of testimony, the melting of Arctic sea ice and the opening up of the Northwest and Northeast

Passages (the poor Transpolar Sea Route often continuing to receive short shrift) are cited as contributing factors, even if how one is supposed to lead to the other is unclear.

If retreating sea ice were indeed such a significant potential threat to peace, one might think that a more forward-thinking response to a situation that is unfolding over years and even decades would be to address that causative factor rather than sleepwalk into conflict. But carbon dioxide emissions continue increasing, the world continues warming, and the Arctic continues melting.

There are, of course, alternative approaches. Consider, for example, the Barents Sea. Forming something of a transition zone between the North Atlantic and Arctic Oceans, it is bordered to the south by the Kola Peninsula, on which is located Murmansk; to the west by the Norwegian Sea; and to the north by Svalbard and Franz Josef Land. Its eastern edge delineates the maritime boundary between Norway and Russia; but for decades, the exact location of that eastern edge was a matter of dispute seemingly without resolution. Norway first requested discussions on establishing a boundary in 1957, which the Soviet Union rejected; formal negotiations did take place in 1974 but could not reach a definitive conclusion to everyone's satisfaction, leaving a "gray zone" that both countries claimed. Each side agreed to work cooperatively on setting fisheries quotas in the disputed area and not to conduct any hydrocarbon or mineral exploration or mining there, but the boundary—rather like the disagreement between Canada and the United States as to whether the Northwest Passage is an international right-of-way or internal waters—seemed destined to be held in permanent abeyance, on an "agree to disagree" basis. Then, to the surprise of just about everyone, the two countries' foreign ministers announced in 2010 that they had reached agreement on a boundary that essentially bifurcated the gray zone, each country giving up approximately 34,000 square miles of claim to settle the dispute. And while both the 2014 Russian annexation of Crimea and the subsequent full invasion of Ukraine have soured relations between the two countries and resulted in Russia's exclusion from most Arctic forums, Oslo and Moscow have continued working quietly behind the scenes on managing the Barents Sea, for example agreeing in October 2023 on a fisheries quota for the 2024 season.

And then there is Hans Island.

Three hundred and twenty acres in area and known in Inuktitut and Greenlandic as Tartupaluk, or "kidney," after its shape, Hans Island lies in the middle of the strait that separates Ellesmere Island in Canada from Greenland and lies directly on the agreed international boundary between the two. When the border was agreed in 1973, both sides claimed it and neither side was willing to cede ground. In 1984, the Danish minister for Greenland visited the island, planted the Danish flag, and left a message that, translated to English, read, "Welcome to the Danish island." He also left a bottle of liquor, which was initially reported as being brandy but was in fact seemingly Schnapps—specifically, it is said, a bottle of Gammel Dansk, which translates as "Old Danish." That same year, Canadian soldiers visited, planted a Canadian flag, and left a bottle of Canadian Club whisky.

Occasionally, the friendly dispute threatened to become something more serious. In the early years of the twenty-first century, Denmark twice dispatched frigates with soldiers to the island, prompting the predictably fevered response from some Canadian politicians who saw it as an affront to their nation's Arctic aspirations.

"Denmark's soldiers land on Canadian Arctic territory, hoist their flag, claim the island as their own and Canada does nothing," fumed one Conservative Party lawmaker in 2004, as quoted by Amanda Coletta in the *Washington Post* two decades later.

Despite such occasional outbreaks of Sturm und Drang, the so-called Whisky War continued in a largely peaceable manner, the two sides exchanging notes and liquor until, in 2022, they came to an agreement by which ownership of the island would be split, more or less evenly, along a natural fault line. The agreement created the third-shortest land border between two countries (behind only the contested border between Gibraltar and Spain and a five-hundred-foot boundary between Botswana and Zambia) and gave both Canada and Denmark their second land neighbor.

"I think it was the friendliest of all wars," Mélanie Joly, Canada's foreign minister, told reporters. "I'm happy to see that we're resolving it with friends, partners and allies."

"As we stand here today, we see gross violation of international rules unfold in another part of the world," added Denmark's foreign minister, Jeppe Kofod, in a reference to the recent Ukraine invasion. "In contrast, we have demonstrated how long-standing international disputes can be resolved peacefully and playing by the rules."

In seeking clarity on the future of the Arctic, and particularly the trans-Arctic passages, one finds instead largely only opacity.

The Canadian government was unsure about its Arctic territories, then it embraced them, then it neglected them, and then it made them—and particularly the Northwest Passage—central to the country's identity as an Arctic nation. Then, after some provocative statements and position papers, it took its foot off the gas a little and settled into a more cooperative and less rambunctious approach. The United States is happy to agree to disagree over the status of the Northwest and Northeast Passages but wants to buy Greenland, until it doesn't, and sees a melting Arctic as a security risk that requires significant military investment. The Northwest Passage is open except when it isn't, and actually there are several Northwest Passages, some of which are too shallow for commercial traffic, some of which are too narrow, and some of which are both. But anyway, its convoluted nature, combined with the lack of infrastructure, means it is unlikely to be much of a commercial artery, for all the history and discord surrounding it—unless one day it is. Russia wants international traffic on the Northern Sea Route but erects multiple barriers to its easy establishment; it engages in provocative acts such as dropping a flag at the North Pole and building up a military presence in its Arctic regions but insists it is acting in accordance with established international regimes. And then it behaves outside the Arctic in a manner entirely contrary to accepted norms, resulting in its exclusion from those Arctic regimes and making the success of the NSR more uncertain than perhaps ever before. China is sufficiently interested in the Arctic to demand a seat at the table, to develop its own icebreakers, and to include open polar passages as part of its putative global network of highways, byways, and trade routes; but it isn't so interested that it is ready to devote more than the occasional line to it in its Five-Year Plans—unless, perhaps, you're Chinese, in which case the message about the Arctic you hear from the government is entirely different from the one the government conveys to the rest of the world. The melting of sea ice means that traffic on the Northwest and Northeast Passages is increasing rapidly, but the presence of sea ice means that that traffic is limited in scope and confined to a small window each year.

The Arctic, in other words, is, like the rest of the world—like much of life—messy, complicated, and nuanced. What can be stated with clarity is that, at an uncertain rate over an indefinable period that may be only a decade or less but might be several, sea ice will, absent sufficient corrective action, continue to retreat in the Arctic until it has all but disappeared in the summer months. In so doing, it will make the Northwest, Northeast, and, ultimately, Transpolar Passages become more open to greater amounts of shipping traffic for longer periods of time. Beyond that, much as one might like the Arctic Magic 8 Ball to state, "It is certain," its most honest response would be a repeated "Ask again later."

———— ∞ ————

"I think that maybe around ten or fifteen years ago, there was a ton of excitement around the opening of the Arctic shipping passages, particularly the Northern Sea Route," Mia Bennett of the University of Washington told me. "I think a lot of that was driven on the one hand by growing recognition of climate changes' impacts in the increasing season for shipping in summertime via the northern sea but also by the idea that China was booming, and then there would be interest, particularly from Asian and Chinese shipping companies, about using the Northern Sea Route to link up with markets and Europe. But I think people are just quite wary of dealing with Russia at the moment, so I think geopolitics has kind of gotten in the way, so to speak, of the fact that there is less and less ice with every passing year on average."

Bennett, who publishes the *Cryopolitics* blog and has written at length about Chinese interests in the Arctic and about the Transpolar Sea Route, among other Arctic issues, sees a possible cooling in Beijing's interest as a result of Russia's military adventurism and the subsequent suite of international economic sanctions. But, she argues, while it may have taken its foot off the gas somewhat, it still sees the Arctic as an important element of its continued growth as a global great power.

"Since the 2010s or so," she continued, "as the Arctic rose in prominence globally, I think China, wanting to be seen as a modern state by other states around the world, felt like it needed to have a prestigious program in places like the polar regions, which have typically been the preserve of more industrialized, more modernized Western states. China wanted to get in on the game, too, and be seen as an equal. So, it's a little bit about identity building as a modern cutting-edge nation-state. But China does see in the future that there will be economic potential in the Arctic, whether it's for shipping or minerals extraction or what have you."

Asked to look into her crystal ball and imagine the state of the Arctic thirty, forty, or fifty years hence, however, Bennett demurred.

"I generally shy away from making such a long-term prediction," she said. "I mean, that would put us at 2065, 2075. It's so hard to foresee. . . . Just looking back over the last five years, would anyone have predicted the COVID-19 pandemic? So, I think it's really complicated. I am part of a research project that is looking at how the Arctic will develop by 2050. In one workshop, we brought together a number of Arctic experts and asked them to come up with some scenarios for how the region would look, and they were so divergent from one another. There are people who imagine a future Arctic where ships are crisscrossing the region but they're totally automated, AI [artificial intelligence] is really important, and internet covers the whole region because of Starlink, and things like that. Others pictured a region full of people, with Russia having come back into the Western fold, and tourism completely booming. There are so many different scenarios, I can't really put my finger on one. The only given is that there's going to be dramatically less sea ice, and Indigenous people's ways of life will be really different from what they have been for the past several decades, if not centuries. I think the ecological changes are going to be devastating. That's really the only aspect that I would put money on, so to speak."

During our conversation, Bennett noted that when it came to predicting the future in the Arctic, one of the major stakeholders is working on a different timescale from the others, a theme that Ólafur Ragnar Grímsson has also emphasized. President of Iceland from 1996 to 2016, he has continued to be at the forefront of Arctic affairs as chair of the Arctic Circle, a nonprofit organization whose annual assemblies have become essential events for those interested in the geopolitics of the northern polar region. Taking place over three October days in Reykjavík's beautiful Harpa Conference Centre, the Arctic Circle Assemblies attract several thousand diplomats, activists, journalists, scientists, explorers, military representatives, and heads of state each year. In addition to multiple simultaneous symposia, daily highlights include an afternoon plenary in which Grímsson holds court on stage, conversing with guests ranging from Alaska senators to Inuit leaders to Emirati sultans. Sessions at the assembly can stretch well into the night, and even in his eighties, Grímsson is energetically engaged throughout the proceedings. In a quieter moment when I was back home in Vermont, it took little encouragement for him to lead me on a rhetorical journey through the Arctic's possible futures as we spoke on the phone.

"It's very important to have a historical perspective on this and to realize that these things do not happen quickly," he began. "Although it is worth remembering what Jiang Zemin, the former president of China, said to me more than twenty years ago when he said, 'The problem with you in the West is you don't understand that for us in China, fifty years is not a long time.' So, it depends on which historical perspective you are using: the Chinese one or the American, Western perspective, where economic progress is measured in annual quarters."

One should also recognize, he reminded me, that the Arctic truly being sufficiently ice-free to permit even relatively limited commercial shipping through some of its more northerly reaches is an extremely recent phenomenon. After all, he noted, it has been only a decade or so since the *Xue Long* became the first Chinese vessel to transit the length of the Northern Sea Route. So a great many elements have yet to play out.

"There have been various entities in those ten years who have spent some time and effort and money and so on, examining these future projects and looking at how would you deal with the possibility of these northern sea routes opening up. And most of this attention, most of these plans, have been focused on the Northeast Passage. If you look at the Northwest Passage, there has been much less attention. And part of that is because the most powerful country in the world, the United States, doesn't have a proper harbor up in Alaska.

"So that leaves the third option, which I usually call the center route, using the path directly across the North Pole during the ice-free months, whether they are three or four or five. And what surprised me six, seven, eight years ago, when I had various discussions with Chinese entities, was that they were very interested in the center. And when they were really pressed as to why, they conceded it was that despite the friendly relationship between Russia and China, China would prefer not to rely on Russia regulations or Russian icebreakers, or on going through Russian waters along the Northeast Passage, and much more would prefer to travel straight across the North Pole. But they are the only entity that I know of who have looked at that route because in the Arctic states, there is very little interest in it."

Grímsson is, like Bennett, reluctant to make too precise a prediction of how the Arctic generally, and its putative shipping routes more specifically, might look even in the relatively near future.

"What will happen in the next five or ten or fifteen years, nobody really knows," he said. "If the war continues or the western blockade on Russia continues, I cannot see any European or Western shipping company using the Northern Sea Route. As for the Northwest Passage, until any infrastructure is built, no company in its right mind would use the Northwest Passage. And if the Americans and Canadians don't start building infrastructure, then the Northwest Passage will be out. That's clear. The only game in town will be either the Northern Sea Route or the polar route for however many months each year it will be ice-free. And if you look at the market, there are not many nations that have the capabilities to build the vessels of the necessary strength to sail these routes.

"And that's where we are."

The sheer number of people at an Arctic Circle Assembly never fails to impress, nor does the sheer diversity of topics: from food security to Indigenous languages to the Arctic policy of the European Union. It is unsurprising to see large delegations from all eight Arctic nations, plus near-Arctic countries such as the Baltic states and Great Britain. At first blush, it feels incongruous that one of the best-represented countries is South Korea, until you appreciate that the country has strong research programs at both ends of the world and, as a major shipping nation, is deeply invested in any potential new transit routes. It is a reminder of how the fate of the Arctic extends far beyond the region's boundaries.

My first assembly, in 2022, was at times an unexpectedly fraught one. It was the first since Russia invaded Ukraine, leading to its exclusion from most regional forums—even while it was the chair of the largest one, Arctic Council. A palpable anxiety pervaded several of the sessions: from Baltic and Scandinavian nationals genuinely concerned about what Moscow's aggression meant for the rest of its neighbors, and from most of those gathered about what the situation portended for cooperation and collaboration in the Arctic. In the absence of any Russian attendees, some of those from China took it upon themselves to be the ones to push gently back against what they saw as overly pro-Western narratives, the final day's crescendo being what might be described in diplomatic circles as a frank exchange of ideas between NATO admiral Rob Bauer and the Chinese ambassador to Iceland, He Rulong.

Unsurprisingly, given the broader context, there was no shortage of panels and presentations on issues of geopolitics and global and regional security. Among the speakers at one of those was Marisol Maddox, the senior Arctic analyst at the Polar Institute of the Woodrow Wilson International Center for Scholars in Washington, DC. There is little to nothing in these pages with which she isn't already innately familiar, given that her work encompasses essentially the nexus of the Arctic, climate change, security, and geopolitics.

When we talked at length a while later, one of her first enjoinders was not to forget that the future of the Arctic, and the viability of the passages, does not depend only on geopolitical and geographic considerations in the region. While cost-benefit calculations presently come down strongly against sending cargo through the Northwest Passage or Northern Sea Route, that equation could change, not only if sea ice continues to decrease but also if, for example, increased instability and conflict in the Red Sea region were to lead to greater piracy or otherwise threaten the security of shipping through the Suez Canal or around the Cape of Good Hope. And climate change affects more than the Arctic: Panama, for example, is at time of writing suffering through a drought that began in early 2023, one consequence of which has been greatly reduced rainfall—in October 2023, the amount of rain was 43 percent less than average, making it the driest October since the 1950s. That has led to a decrease in water levels in Gatun Lake, which holds the water supply for the Panama Canal: in January 2024, the lake was six inches lower than it had been a year before. Lower water levels in the canal limit the number and size of vessels that can pass through, and by November 2023, Panama had capped the number of allowable daily transits at twenty-four, down from the usual thirty-eight. That's still far more traffic than transits the Northern Sea Route on a daily basis, but over the course of a year, it could lead to four thousand fewer vessels passing through one of the world's primary shipping routes and thus looking for an alternative route. As Maddox observed, "What's happening in other parts of the world affects the risk calculus" around the Arctic passages.

There is another kind of risk calculus, namely, the potential for an ecological disaster as a result of routing cargo or container vessels through Arctic waters. Environmental concerns were behind the 2017 adoption by the International Maritime Organization of the Polar Code, which mandates that vessels operating in either the Arctic or the Antarctic be constructed to specific strengths and which sets standards for crew training, ship maintenance, equipment and clothing onboard, and the type of lifeboats that are carried. As of July 1, 2024, it also requires that some vessels operating in polar regions not use or carry heavy fuel oils (HFOs), which are viscous and high in sulfur. The ban was introduced partly to address toxic pollution from both sulfur and black carbon, soot particles that not only are hazardous to human health but also magnify warming in the Arctic by settling on ice and snow and increasing the amount of solar radiation they absorb. But HFOs also break down extremely slowly in polar waters. When combined with the relatively paucity of infrastructure and search and rescue capabilities, this means that a spill would likely prove almost impossible to clean up.

Unfortunately, the ban at present applies only to ships flying the flags of non-Arctic states; Arctic nations can waive the requirements until January 2029, meaning that an estimated 74 percent of ships that use HFOs in the Arctic will be able to continue doing so—although Canada, for example, issued a domestic ban through an interim order when the rule came into effect, and Norway has operated a strict ban on HFOs in ships in Svalbard waters since January 2022. Additionally, there are concerns that some of the lower-sulfur-content fuels that shipping companies are adopting instead may not be significantly less polluting in Arctic waters than HFOs.

And no matter what kind of regulations may exist on paper, the involvement of humans means that accidents are always possible, perhaps even likely, not just with the ships themselves but also with the associated infrastructure. Maddox worries that as Russia isolates itself, those risks will only increase.

"They've essentially limited who is willing to work with them," she noted. "So they're now much more dependent upon China. But now Chinese companies are also concerned about exposure to sanctions and secondary sanctions. There was recently an announcement about a Chinese company that had been drafted in to work on the Yamal LNG 2 facility—to replace a Western company—getting cold feet because of sanctions fears and saying, 'We're just not going to do business with Russia.' Unfortunately, that just makes it so that if Russia is desperate to develop the infrastructure, they'll work with whoever just to get it done, and that heightens the risk that it's going to be done in a shoddy way, which then increases the risk that there's going to be some type of failure or serious environmental incident." At the same time, Moscow is turning to its friends and vassals for investment and materials to export: after Belarus's 2020 elections were widely condemned for clear fraudulence, Russia was almost alone in sticking by the side of Belarusian president Alexander Lukashenko, part payment for which is Minsk investing in port infrastructure in Murmansk. And with Belarus also subject to sanctions and with Ukraine lying between it and Russia's Black Sea ports, the Northern Sea Route provides Belarus with an export pathway for its potash industry, which in turn adds to the amount of cargo that Moscow can boast about being carried via the NSR.

The broader issue, however, remains that all such investment takes for granted the fact that the Northeast Passage—and the Northwest Passage, and ultimately the Transpolar Sea Route—will become sufficiently ice-free to enable commercial shipping to transit them with greater regularity.

And while there is an element of inevitability about that, given that, as we saw earlier, melting sea ice begets more melting sea ice, it is hard to escape the feeling that far more effort is being deployed in preparing for that eventuality than there is in attempting to preempt it and bring warming to a halt.

"My perception of it all is that climate change is the adversary that nobody wants to deal with," Maddox said with a sigh. "It's an actorless threat, and we're just not cut out to deal with those. That's not the adversary we want. We want there to be a battle. We want to go kick butt. Military guys and gals love that, right? Bad guys with weapons are the adversaries they're made for.

"But we have to look at root causes—and in a way, climate change isn't actually the problem. The problem is that our society has evolved in a way that's not conducive to longevity. We have to have an introspective process that enables us to shift to a different way of interacting with the world. And that's complicated. Culture and human behavior and related issues: they are total headaches that nobody wants to deal with. That's what we have to do, but nobody wants to hear that. And that's the problem."

———

There are those who may take issue with the notion that little to nothing is being done to address climate change. After all, didn't the countries of the world agree in Paris in 2015 to work to limit warming to no more than 2°C (3.6°F) above preindustrial levels, and even to strive to restrict warming to 1.5°C (2.7°F)? Is it not also true that a number of countries—including major emitters such as China—are on course to see their carbon emissions peak by 2030 or even earlier, while the United States has apparently even passed its peak emissions years? Yes, that is all correct, but actions, of course, speak louder than words. In his July 2024 address to the United States explaining why he had decided not to seek reelection, President Joe Biden referred to climate change as an existential threat, but few countries' actions are indicative of their regarding it as such. Biden himself can at least claim credit for the Inflation Reduction Act of 2023, which provided investment for clean energy and encouraged private-sector divestment from fossil fuels; but one-half of the country's body politic has been consumed by insanity, including but not restricted to a refusal to even acknowledge the reality of climate change. The world's two most populous nations, China and India, are investing in coal plants even as India also is developing renewable energy programs and China leads the world in solar photovoltaics and electric cars.

Overall levels of carbon dioxide in the atmosphere continue to rise and so, too, do global temperatures, so much so that Earth is even now bumping up against that 1.5°C target. Projecting future emissions scenarios is a contentious topic, depending on whether one falls on the optimistic or pessimistic side of the ledger, but a United Nations analysis released in late 2023 argued that countries' current pledges under the Paris Agreement to reduce their emissions would still put Earth on course for nearly 3°C (5.2°F) of warming this century. In order to keep warming to 1.5°C, said the authors of the United Nations Environment Programme's *Emissions Gap Report 2023*, greenhouse gas emissions globally would need to fall by 42 percent by 2030; instead, they are on course to increase by 3 percent—an improvement on the projected 16 percent increase at the time the Paris Agreement was signed, but still far from what is needed.

In a 2022 paper in the journal *Anthropocene*, a team of researchers led by Jinlei Chen of the Chinese Academy of Sciences projected changes in sea ice and the navigability of the Arctic passages with global warming of 2°C and 3°C. It concluded that with 2°C of warming, open-water vessels—those not ice strengthened—would remain unable to navigate either the Northwest or Northeast Passage for most of the year, although parts of both routes might be accessible in September. For Polar Class 6 (PC 6) vessels—among the very lightest of ice classes—the bulk of both passages would most likely be accessible from October through December. With warming of 3°C, even the most recalcitrant and ice-clogged areas, such as the northern limits of the Canadian Arctic Archipelago—would yield to both types of vessels for several months per year and, in the case of PC 6 vessels, for most months.

Other studies have shown that the path to a truly ice-free and navigable Arctic will likely come in fits and starts, however. A 2024 study published in *Nature* found that as younger sea ice in the Northwest Passage melts, it no longer acts as a barrier to the older, more persistent ice that resides high in the archipelago; as that ice is released, it actually creates barriers and choke points along the Passage, not only limiting the potential for future commercial transits but also impeding essential supply routes for communities along the Passage. And, as ice melts and exposes the surface of ocean to the air, that water and the air above it exchange moisture, leading to fog. Already an issue in parts of the Northern Sea Route, it is projected to become severe enough along parts of the Northwest Passage that it could slow projected shipping times by as much as three days and increase the likelihood of collisions with icebergs and thick floes.

That reflects the observations of what those on the ground—or, if you wish, on the water—are reporting.

"We're seeing multiyear ice in the southern waters [of the Canadian Arctic Archipelago] because it's not being held back by ice bridges," David "Duke" Snider of Martech Polar Consulting told me. "The same thing is happening over on the Nares Strait between Greenland and the Canadian Archipelago with big icebergs that used to be held in place for several years at a time. It isn't just glacial ice; we're seeing multiyear sea ice sliding down into northern Baffin Bay when it didn't before. We're even seeing glaciers calving off into the Beaufort Sea; previously, we would never see icebergs in the Beaufort or Chukchi Seas."

Snider pointed out that what is classified as ice-free or nearly ice-free by researchers looking at satellite imagery doesn't necessarily translate to smooth sailing for those who are actually trying to navigate the passages; what might, from on high, seem insignificant residues of ice can be significant for a ship attempting to make its way through a narrow channel. That said, Snider agreed there is no doubt whatsoever about the direction in which the environment is heading.

"The bottom line is that climate change is very clearly affecting the ice situation at both poles," he said. "There is no doubt that there is less multiyear ice year-in, year-out in the Arctic than when I had hair on the top of my head. We see it. The seasons are changing. They're lengthening. Interestingly, we're not seeing a huge change at the beginning of the season. The big change is at the end of the season; the return to freezing is occurring later and later."

Mark Serreze of the National Snow and Ice Data Center pointed out that attempting to define when one or more of the passages might become truly navigable is, in many respects, the wrong metric: to a large extent, the Northern Sea Route is open for much of the year already simply because Russia has built a fleet of nuclear-powered icebreakers that can crunch through even the thickest ice. He, too, pointed to the difficulties of determining future sea ice extent with any precision, even if the trend is crystal clear.

"In the science community, there's all this talk about 'When do you lose summer sea ice?' There are all kinds of estimates. I'm on record as saying it could be as early as 2030, for which I've been vilified on the internet quite a bit these days. The issue, of course, is that there will still be winter ice for a long time. Even if you have a good, clear transpolar route, for example, you'll still get winter ice. But when will it be clear in the summer? You could be looking at a decade from now; it could be several decades from now. There are so many unknowns." Serreze noted, for example, that after the huge collapses in Arctic sea ice extent in 2007 and 2012, summer sea ice extent has remained low but without much variation. Does that mean another sudden plunge is just around the corner, one that, as in those previous two years, will cause a rapid reassessment of how quickly ice could disappear?

"In terms of the sea ice of the Arctic, and climate warming in general, we see the future kind of like painting with a broad brush," said Serreze. "We know what's out there. It's going to get warmer. We're going to lose sea ice. That's already happening. But the details of just when you lose the sea ice, or how rapidly the Greenland ice sheet will melt down? There are still so many unknowns out there. So, in other words, the devil is in the details. But of course, it's the details that often matter the most."

A 3°C rise in global temperatures will, of course, affect far more than just the Arctic. At that point, scientists predict, the world could pass several catastrophic points of no return, with consequences including the displacement of over a billion people; the collapse of ice caps, leading to uncontrollable sea level rise; frequent and devastating extreme weather events; and the endangerment of the Amazon and Congo River Basins. In such circumstances, it is hard to imagine that society's focus will be on being able to receive their mail-order goods a day or two earlier by shipping them through the Arctic. As I write this, wildfires are spreading across California and have devastated Jasper, Alberta—the latest in a string of extreme weather events that have torn through different parts of the planet in the past several years. And that is with just under 1.5°C of warming. Three degrees is barely imaginable and not a world in which many of us would care to live. And yet, when we anticipate a time when ships can sail unimpeded through the Arctic passages and finally achieve the dream for which Sir John Franklin and his crew gave their lives, that is the world for which we are preparing and that we are accepting as a given.

There are multiple reasons why people may not embrace action to combat climate change. A major part is that they find the prescribed steps to counteract it threatening to their ways of life and belief systems. Others choose, for similar reasons, not to believe that climate change is even happening. And for still others, of course, it is simply easier and more profitable to pursue the present course. Even among those who recognize the validity of the issue, there can be a sense of urgency deferred: when judged by the increasingly short attention spans of modern society, climate change is unspooling at a slow pace. Like the proverbial frog in a slowly boiling pot of water, there is a tendency to become all too easily acclimated to a "new normal," even as that normal changes from one day to the next. It is perhaps why climate change's habit of spawning extreme weather events, which are both tangible and frequent, is the feature of a warming planet that provokes the greatest agitation and foreboding and is the link to a changing climate that those who profit from casting doubt on global warming's existence work hardest to discredit.

When it comes to climate change in the Arctic, there is another issue: the fact that for the vast majority of Earth's population, the Arctic remains both a hostile wasteland and a blank canvas, a place that has existed in the human imagination solely to be explored and exploited. If there is a belated recognition that the region does have its own permanent inhabitants, their collective bargaining power is diminished by their paucity: the total population north of the Arctic Circle is barely 10 percent of that of metropolitan Tokyo. Their wishes will not be the determining factor in the future of the Arctic over the coming decades.

The reality is that, ultimately, the fate of the Arctic will not be decided in the Arctic, by those who live there or even those who govern it. Those involved can make all the plans they want about icebreaker fleets, shipping routes, and open-water passages; but if those plans come to fruition, it will be because the world collectively has decided to acquiesce in the face of greenhouse gas emissions and rising temperatures, and of the obduracy and grift of those who have chosen to deny the overwhelming evidence and obstruct the measures that need to be taken to turn down the thermostat on an overheating planet.

If we do enter a time when container traffic is routinely traversing an ice-free Arctic Ocean, or the Northern Sea Route truly is a commercially viable shortcut for shipping, it will be far from the only fundamental change to the world as we know it. If, instead, the Arctic in the twenty-second century looks quite similar to that of the twenty-first or even the twentieth, it will be because the world as a whole has slammed on the brakes before it collectively careens over the cliff. The Arctic is not a remote, inhospitable realm far from the concerns of most of humanity. It is an integral and important element of our planetary makeup. Its fate and that of the rest of the planet are interlinked; its future and our future are inseparable.

———∞———

Sources

Beginnings

Cordell, Jake. "Could Russia Benefit from the Suez Canal Blockage?" *Moscow Times*, March 26, 2021. https://www.themoscowtimes.com/2021/03/26/could-russia-benefit-from-the-suez-canal-blockage-a73385.

Danilov, Peter B. "Northern Sea Route Transit Traffic Remains Modest." *High North News*, February 12, 2021. https://www.highnorthnews.com/en/northern-sea-route-transit-traffic-remains-modest.

Dickens, Charles. "The Lost Arctic Voyagers." *Household Words*, December 2 and 9, 1854. https://victorianweb.org/authors/dickens/arctic/pva342.html.

Fernández-Armesto, Felipe. *Pathfinders: A Global History of Exploration*. New York: W. W. Norton, 2006.

Freedman, Andrew. "Global Warming Has Profoundly Transformed Arctic in Just 15 Years, Report Warns." *Washington Post*, December 8, 2020. https://www.washingtonpost.com/weather/2020/12/08/arctic-climate-change-report-siberia/.

Higgins, Jenny. "The *Matthew*." Newfoundland and Labrador Heritage, 2013. https://www.heritage.nf.ca/articles/exploration/matthew.php.

Hutton, Samuel King. *Among the Eskimos of Labrador: A Record of Five Years' Close Intercourse with the Eskimo Tribes of Labrador*. Toronto: Musson Book Company, 1912.

Jones, Evan T., and Margaret M. Condon. *Cabot and Bristol's Age of Discovery: The Bristol Discovery Voyages 1480–1508*. Bristol, UK: University of Bristol, 2016.

Pompeo, Michael R. "Looking North: Sharpening America's Arctic Focus." Speech given May 6, 2019, Rovaniemi, Finland. https://2017-2021.state.gov/looking-north-sharpening-americas-arctic-focus.

Rayess, Rami. "Dead in the Water: Northwest Passage Will Replace Suez as World's Transport Route." Al Arabiya, September 14, 2021. https://english.alarabiya.net/views/2021/09/14/Dead-in-the-water-Northwest-Passage-will-replace-Suez-as-world-s-transport-route.

Savage, Luiza Ch. "How Russia and China Are Preparing to Exploit a Warming Planet." *Politico*, August 29, 2019. https://www.politico.com/story/2019/08/29/russia-china-climate-change-1691698a.

Shabecoff, Philip. "Global Warming Has Begun, Expert Tells Senate." *New York Times*, June 24, 1988. https://www.nytimes.com/1988/06/24/us/global-warming-has-begun-expert-tells-senate.html.

Sheppard, Kate. "A Voice in the Wilderness." *Guardian* (US edition), June 23, 2008. https://www.theguardian.com/commentisfree/2008/jun/23/climatechange.carbonemissions2.

Stefansson, Vilhjalmur. *The Friendly Arctic: The Story of Five Years in Polar Regions*. New York: Macmillan, 1921.

Yee, Vivian, and James Glanz. "How One of the World's Biggest Ships Jammed the Suez Canal." *New York Times*, July 17, 2021, updated July 19, 2021. https://www.nytimes.com/2021/07/17/world/middleeast/suez-canal-stuck-ship-ever-given.html.

Northwest

Adams, Peter, and Maxwell J. Dunbar. "Arctic Archipelago." *Canadian Encyclopedia*, March 9, 2006. https://www.thecanadianencyclopedia.ca/en/article/arctic-archipelago.

Aporta, Claudio. "Shifting Perspectives on Shifting Ice: Documenting and Representing Inuit Use of the Sea Ice." *Canadian Geographer* 55, no. 1 (2011): 6–19. https://doi.org/10.1111/j.1541-0064.2010.00340.x.

Aporta, Claudio, and Charlie Watt. "Arctic Waters as Inuit Homeland." Chap. 12 in *Routledge Handbook of Indigenous Peoples in the Arctic*, edited by Timo Koivurova, Else Grete Broderstad, Dorothée Cambou, Dalee Dorough, and Florian Stammler. London: Routledge, 2021.

Boutilier, Alex. "'We're Going to Find' Franklin Expedition, Stephen Harper Vows." *Toronto Star*, August 25, 2014. https://www.thestar.com/news/canada/we-re-going-to-find-franklin-expedition-stephen-harper-vows/article_50e1de90-6cec-512b-a87c-6bf387c1749b.html.

Cecco, Leyland. "Mike Pompeo Rejects Canada's Claims to Northwest Passage as 'Illegitimate.'" *Guardian* (US edition), May

7, 2019. https://www.theguardian.com/us-news/2019/may/07/mike-pompeo-canada-northwest-passage-illegitimate.

Demuth, Bathsheba. "What Made the Thule Move? Climate and Culture in the High Arctic." HistoricalClimatology.com, September 12, 2016. https://www.historicalclimatology.com/features/what-made-the-thule-move-climate-and-culture-in-the-high-arctic.

Fortin, M. A. "The Harper Conservatives' Arctic Policy: Did It Really Make a Difference?" Canadian Forces College, JCSP 42, 2015–2016. https://www.cfc.forces.gc.ca/259/290/318/305/fortin.pdf.

Fram—the Polar Expedition Museum. "Sverdrup, Otto Neumann Knoph (1854–1930)," n.d. https://frammuseum.no/polar-history/explorers/otto-neumann-knoph-sverdrup-1854-1930/.

Inuit Circumpolar Council (ICC) Canada. "The Sea Ice Is Our Highway: An Inuit Perspective on Transportation in the Arctic." March 2008. https://www.inuitcircumpolar.com/wp-content/uploads/2019/01/20080423_iccamsa_finalpdfprint.pdf.

Inuit Circumpolar Council (ICC) Canada. "The Sea Ice Never Stops: Circumpolar Inuit Reflections on Sea Ice Use and Shipping in Inuit Nunaat." December 2014. https://www.inuitcircumpolar.com/wp-content/uploads/Sea-Ice-Never-Stops-Final.pdf.

Kikkert, Peter. "Canadian Arctic Expedition." *Canadian Encyclopedia*, February 6, 2006. http://www.thecanadianencyclopedia.ca/en/article/canadian-arctic-expedition.

Kikkert, Peter, and P. Whitney Lackenbauer. "'On Hallowed Ground': *St. Roch*, Sovereignty, and the 1944 Northwest Passage Transit." *Northern Mariner* 29, no. 3 (Fall 2019): 213–32. https://doi.org/10.25071/2561-5467.212, https://tnm.journals.yorku.ca/index.php/default/article/view/212/201.

Marcus, Alan R. *Out in the Cold: The Legacy of Canada's Inuit Relocation Experiment in the High Arctic*. Copenhagen: International Work Group for Indigenous Affairs, 1992.

Marsh, James H. "Baffin Island." *Canadian Encyclopedia*, February 6, 2006. https://www.thecanadianencyclopedia.ca/en/article/baffin-island.

Marsh, James H. "Ellesmere Island." *Canadian Encyclopedia*, April 15, 2012. https://www.thecanadianencyclopedia.ca/en/article/ellesmere-island.

McDiarmid, Margo. "Stephen Harper and the Obsession with Franklin." CBC News, September 3, 2014. https://www.cbc.ca/news/politics/stephen-harper-and-the-obsession-with-franklin-1.2754180.

Morrison, W. R. "Canadian Arctic Sovereignty." *Canadian Encyclopedia*, February 6, 2006. https://www.thecanadianencyclopedia.ca/en/article/arctic-sovereignty.

Mulvaney, Kieran. "'If It Gets Me, It Gets Me': The Town where Residents Live Alongside Polar Bears." *Guardian* (US edition), February 13, 2019. https://www.theguardian.com/world/2019/feb/13/churchill-canada-polar-bear-capital.

Mulvaney, Kieran. "Welcome to the Polar Bear Capital of the World: 'It's Kind of Epic.'" *National Geographic*, December 22, 2022. https://www.nationalgeographic.com/animals/article/polar-bear-capital-of-the-world-hudson-bay-churchill.

Noack, Rick, John Wagner, and Felicia Somnez. "Trump Attacks Danish Prime Minister for Her 'Nasty' Comments about His Interest in U.S. Purchase of Greenland." *Washington Post*, August 21, 2019. https://www.washingtonpost.com/world/europe/danes-furious-over-postponement-of-trumps-visit-call-his-behavior-insulting/2019/08/21/82d0a5f4-c3b8-11e9-8bf7-cde2d9e09055_story.html.

Pelly, David F. "Dundas Harbour: Keeping Watch over the Northwest Passage." *Above & Beyond: Canada's Arctic Journal* (July/August 2013): 15–18. https://www.davidpelly.com/resources/Dundas-Hbr.pdf.

Plouffe, Joël. "Stephen Harper's Arctic Paradox." Canadian Global Affairs Institute, December 2014. https://www.cgai.ca/stephen_harpers_arctic_paradox.

Rennie, Steve. "Franklin Find as Much about Sovereignty as Solving a Mystery." CBC News, September 11, 2014. https://www.cbc.ca/news/canada/north/franklin-find-as-much-about-sovereignty-as-solving-a-mystery-1.2763117.

Sloan, Gene. "A History of Cruises through the Northwest Passage." *USA Today*, October 7, 2016. https://www.usatoday.com/story/travel/cruises/2016/10/07/northwest-passage-cruise-history/91720028/.

Smith, Gordon W. "The Transfer of Arctic Territories from Great Britain to Canada in 1880, and Some Related Matters, as Seen in Official Correspondence." *Arctic* 14, no. 1 (1961): 53–73. https://doi.org/10.14430/arctic3660, https://journalhosting.ucalgary.ca/index.php/arctic/article/view/66710/50623.

United Press International. "First Passenger Ship Navigates Northwest Passage." September 12, 1984. https://www.upi.com/Archives/1984/09/12/First-passenger-ship-navigates-Northwest-Passage/8564463809600/.

Woolf, Nicky. "Canada Uses Franklin Expedition Wreck to Boost North-West Passage Claim." *Guardian* (US edition), September 13, 2014. https://www.theguardian.com/world/2014/sep/13/canada-uses-franklin-expedition-wreck-north-west-passage-claim.

Northeast

American Bureau of Shipping. "Navigating the Northern Sea Route: Status and Guidance," n.d. https://ww2.eagle.org/content/dam/eagle/advisories-and-debriefs/ABS_NSR_Advisory.pdf.

Bennett, Mia. "Northern Lights to Neon Lights: Kirkenes to Transform into Polar Chinatown." *Cryopolitics* (blog), November 8, 2018. https://www.cryopolitics.com/2018/11/08/kirkenes-polar-chinatown/.

Bennetts, Marc. "Russian Islands Declare Emergency after Mass Invasion of Polar Bears." *Guardian* (US edition), February 11, 2019. https://www.theguardian.com/world/2019/feb/11/russian-islands-emergency-mass-invasion-polar-bears-novaya-zemlya.

Borshoff, Isabella. "Norway's 'Northernmost Chinatown' Eyes Arctic Opportunity." *Politico*, November 20, 2019. https://www.politico.eu/article/norway-kirkenes-china-influence-arctic-shipping-opportunity/.

Byers, Michael. "The Universal Unimak Pass." *Arctic in Context* (blog). Henry M. Jackson School of International Studies, University of Washington, February 4, 2015. https://jsis.washington.edu/aic/2015/02/04/the-universal-unimak-pass.

Crowdy, Wendy. "Provideniya: An Ominous Introduction to Beringia." *TravelArk 2.0* (blog), August 23, 2018. http://v2.travelark.org/travel-blog-entry/crowdywendy/10/1542325320.

Gunnarsson, Björn, and Arild Moe. "Ten Years of International Shipping on the Northern Sea Route: Trends and Challenges." *Arctic Review on Law and Politics* 12 (2021): 4–30. https://doi.org/10.23865/arctic.v12.2614.

Hamilton, Lawrence, Jochen Wirsing, Jessica Brunacini, and Stephanie Pfirman. "Arctic Knowledge of the U.S. Public." Arctic Research Consortium of the United States (ARCUS), May 2017. https://www.arcus.org/witness-the-arctic/2017/5/highlight/2.

Huffines, Eleanor. "Most Large Ships Transiting Arctic Use New Routes That Help Protect Environment and Communities." The Pew Charitable Trusts, May 20, 2020. https://www.pewtrusts.org/en/research-and-analysis/articles/2020/05/20/most-large-ships-transiting-arctic-use-new-routes-that-help-protect-environment-and-communities.

Humpert, Malte. "Chinese Shipping Company COSCO to Send Record Number of Ships through Arctic." *High North News*, June 13, 2019. https://www.highnorthnews.com/en/chinese-shipping-company-cosco-send-record-number-ships-through-arctic.

Japan Association of Marine Safety. *Northern Sea Route Handbook*, 2015. https://www.nikkaibo.or.jp/pdf/NorthernSeaRouteHandbook_E.pdf.

Li, Xiaoyang, Natsuhiko Otsuka, and Lawson W. Brigham. "Spatial and Temporal Variations of Recent Shipping along the Northern Sea Route." *Polar Science* 27 (2021): 100569. https://doi.org/10.1016/j.polar.2020.100569, https://www.sciencedirect.com/science/article/pii/S1873965220300785?via%3Dihub.

Moscow Times. "Russia Scrambles to Escort Ships Stuck in Arctic Shipping Route—Reports." November 22, 2021. https://www.themoscowtimes.com/2021/11/22/russia-scrambles-to-escort-ships-stuck-in-arctic-shipping-route-reports-a75624.

Solski, Jan Jakub. "The Northern Sea Route in the 2010s: Development and Implementation of Relevant Law." *Arctic Review on Law and Politics* 11 (2020): 383–410. https://doi.org/10.23865/arctic.v11.2374.

Staalesen, Atle. "Chinese Shippers Shun Russian Arctic Waters." *Barents Observer*, August 22, 2022. https://thebarentsobserver.com/en/industry-and-energy/2022/08/chinese-shippers-shun-russian-arctic-waters.

Todorov, Andrey. "New Russian Law on Northern Sea Route Navigation: Gathering Arctic Storm or Tempest in a Teapot?" Belfer Center for Science and International Affairs, Harvard Kennedy School, March 9, 2023. https://www.belfercenter.org/publication/new-russian-law-northern-sea-route-navigation-gathering-arctic-storm-or-tempest-teapot.

Zhang, Yiru, Qiang Meng, and Liye Zhang. "Is the Northern Sea Route Attractive to Shipping Companies? Some Insights from Recent Ship Traffic Data." *Marine Policy* 73 (2016): 53–60. https://doi.org/10.1016/j.marpol.2016.07.030. https://www.sciencedirect.com/science/article/abs/pii/S0308597X16302081?via%3Dihub.

North

Bennett, Mia. "The Arctic Shipping Route No One's Talking About." *Cryopolitics* (blog), April 23, 2019. https://www.cryopolitics.com/2019/04/23/transpolar-passage/.

Bennett, Mia. "In Just 20 Years, Ships Could Cross an Open Arctic Ocean." *Cryopolitics* (blog), September 4, 2020. https://www.cryopolitics. com/2020/09/04/sailing-open-arctic-ocean/.

Bennett, Mia M., Scott R. Stephenson, Kang Yang, Michael T. Bravo, and Bert de Jonghe. "The Opening of the Transpolar Sea Route: Logistical, Geopolitical, Environmental, and Socioeconomic Impacts." *Marine Policy* 121 (2020): 104178. https://doi.org/10.1016/j. marpol.2020.104178, https://www.sciencedirect.com/science/article/abs/ pii/S0308597X2030453X.

Chiacchia, Benjamin. "The Case for an Arctic Treaty." *Proceedings of the U.S. Naval Institute* 146, no. 5 (May 2020). https://www.usni.org/ magazines/proceedings/2020/may/case-arctic-treaty.

Cunliffe, Barry. *The Extraordinary Voyage of Pytheas the Greek*. New York: Walker & Company, 2002.

Doshi, Rush, Alexis Dale-Huang, and Gaoqi Zhang. "Northern Expedition: China's Arctic Activities and Ambitions." Brookings Institution, April 2021. https://www.brookings.edu/articles/ northern-expedition-chinas-arctic-activities-and-ambitions/.

Eckel, Mike, Wojtek Grojec, and Ivan Gutterman. "Under Sea, under Stone: How the U.S. Claimed Vast New Arctic Territory—in an Unusual Way." *Radio Free Europe / Radio Liberty*, n.d. https://www.rferl.org/a/arctic- sea-claims-interactive-map/32793427.html.

Eiterjord, Trym. "What the 14th Five-Year Plan Says about China's Arctic Interests." Arctic Institute, November 23, 2023. https://www. thearcticinstitute.org/14th-five-year-plan-chinas-arctic-interests/.

Fischetti, Mark. "Nations Claim Large Overlapping Sections of Arctic Seafloor." *Scientific American*, August 1, 2019. https://www.scientificamerican.com/article/ nations-claim-large-overlapping-sections-of-arctic-seafloor/.

Gorbachev, Mikhail. "Speech in Murmansk at the Ceremonial Meeting on the Occasion of the Presentation of the Order of Lenin and the Gold Star to the City of Murmansk." October 1, 1987. https://www.barentsinfo.fi/ docs/Gorbachev_speech.pdf.

Henriques, Martha. "The Rush to Claim an Undersea Mountain Range." BBC News, July 23, 2020. https://www.bbc.com/future/ article/20200722-the-rush-to-claim-an-undersea-mountain-range.

"The Ilulissat Declaration." May 28, 2008. Adopted at the Arctic Ocean Conference, Ilulissat, Greenland, May 27–29, 2008. https://arcticportal.org/images/stories/pdf/Ilulissat-declaration.pdf.

Irving, Doug. "What Does China's Arctic Presence Mean to the United States?" RAND Corporation, December 29, 2022. https://www.rand.org/pubs/articles/2022/what-does-chinas-arctic-presence-mean-to-the-us.html.

Jonassen, Trine. "'Polarstern' Crew after a Year Trapped in the Sea Ice: 'Nature Is Still Boss in the Arctic.'" *High North News*, November 6, 2020. https://www.highnorthnews.com/en/polarstern-crew-after-year-trapped-sea-ice-nature-still-boss-arctic.

Lamazhapov, Erdem, Iselin Stensdal, and Gørild Heggelund. "China's Polar Silk Road: Long Game or Failed Strategy?" Arctic Institute, November 14, 2023. https://www.thearcticinstitute.org/china-polar-silk-road-long-game-failed-strategy/.

Lanteigne, Marc. "The Rise (and Fall?) of the Polar Silk Road." *The Diplomat*, August 29, 2022. https://thediplomat.com/2022/08/the-rise-and-fall-of-the-polar-silk-road/.

McBride, James, Noah Berman, and Andrew Chatzky. "China's Massive Belt and Road Initiative." Council on Foreign Relations, February 2, 2023. https://www.cfr.org/backgrounder/chinas-massive-belt-and-road-initiative.

McCaskill, Eloise. "Pytheas." *Encyclopedia Arctica*. Vol. 15, *Biographies*, 0625–31, n.d. Dartmouth College Library. https://collections.dartmouth.edu/arctica-beta/html/EA15-57.html.

Mulvaney, Kieran. "How Climate Models Got So Accurate They Earned a Nobel Prize." *National Geographic*, October 5, 2021. https://www.nationalgeographic.com/environment/article/how-climate-models-got-so-accurate-they-earned-a-nobel-prize.

Nakano, Jane, and William Li. "China Launches the Polar Silk Road." Center for Strategic and International Studies, February 2, 2018. https://www.csis.org/analysis/china-launches-polar-silk-road.

Pope, Kristen. "MOSAiC Researchers Adapt to the Challenges of COVID-19 Pandemic and Extreme Arctic Conditions." Yale Climate Connections, May 21, 2020. https://yaleclimateconnections.org/2020/05/mosaic-researchers-adapt-to-challenges-of-covid-19-and-extreme-arctic-conditions/.

Watt-Cloutier, Sheila. "A Message from the Arctic." *Updates from the Field* (blog). Pacific Environment, June 12, 2018. https://www.pacificenvironment.org/a-message-from-the-arctic/.

Weber, Bob. "Canada Pledges to Work with U.S. over Competing Claims to Arctic Sea Floor." CBC News, January 3, 2024. https://www.cbc.ca/amp/1.7073547.

Futures

Austen, Ian. "Canada and Denmark End Their Arctic Whisky War." *New York Times*, June 14, 2022. https://www.nytimes.com/2022/06/14/world/canada/hans-island-ownership-canada-denmark.html.

Baker, Mike. "With Eyes on Russia, the U.S. Military Prepares for an Arctic Future." *New York Times*, March 27, 2022. https://www.nytimes.com/2022/03/27/us/army-alaska-arctic-russia.html.

Burgess, Richard R. "Navy Admirals Detail Russian Arctic Build-Up." *Seapower*, February 15, 2023. https://seapowermagazine.org/navy-admirals-detail-russian-arctic-build-up/.

Chen, Jinlei, Shichang Kang, Qinglong You, Yulan Zhang, and Wentao Du. "Projected Changes in Sea Ice and the Navigability of the Arctic Passages under Global Warming of 2°C and 3°C." *Anthropocene* 40 (2022): 100349. https://doi.org/10.1016/j.ancene.2022.100349, https://www.sciencedirect.com/science/article/abs/pii/S2213305422000303#.

Coletta, Amanda. "Ukraine War Brings Peace—between Canada and Denmark." *Washington Post*, June 14, 2022. https://www.washingtonpost.com/world/2022/06/14/canada-denmark-greenland-hans-island/.

Cook, Alison J., Jackie Dawson, Stephen E. L. Howell, Jean E. Holloway, and Mike Brady. "Sea Ice Choke Points Reduce the Length of the Shipping Season in the Northwest Passage." *Communications Earth & Environment* 5, art. no. 362 (2024). https://doi.org/10.1038/s43247-024-01477-6, https://www.nature.com/articles/s43247-024-01477-6.

Dickie, Gloria. "Climate on Track to Warm by Nearly 3C without Aggressive Actions, UN Report Finds." Reuters, November 20, 2023. https://www.reuters.com/sustainability/climate-energy/climate-track-warm-by-nearly-3c-without-greater-ambition-un-report-2023-11-20/.

Greenwood, Keely. "Sir Michael Palin Unveils Plaque to Franklin's Lost Expedition at Sir John Franklin Pub in Greenhithe." *KentOnline*,

January 23, 2024. https://www.kentonline.co.uk/dartford/news/
sir-michael-palin-pays-emotional-visit-to-kent-pub-300559/.

Gronholt-Pedersen, Jacob, and Gwladys Fouche. "Dark Arctic: NATO Allies
Wake Up to Russian Supremacy in the Region." Reuters, November
16, 2022. https://www.reuters.com/graphics/ARCTIC-SECURITY/
zgvobmblrpd/.

Mathis, Joel. "The New Cold War in the Arctic, Explained." *The Week*,
June 22, 2023. https://theweek.com/climate-change/1024426/
the-new-cold-war-in-the-arctic-explained.

Nilsen, Thomas. "Moscow Hoists Soviet Flags at Svalbard." *Barents
Observer*, June 30, 2024. https://thebarentsobserver.com/en/
arctic/2024/06/back-ussr-soviet-flags-wave-svalbard.

Williams, Holly, and Analisa Novak. "Russia Ramps Up Its Military
Presence in the Arctic Nearly 2 Years into the Ukraine War."
CBS News, December 18, 2023. https://www.cbsnews.com/news/
russia-arctic-military-presence-ukraine-war-nears-two-year-mark/.

Further Reading

Arctic: General, Ecology, and Natural History

Dodds, Klaus, and Mark Nuttall. *The Arctic: What Everyone Needs to Know*. New York: Oxford University Press, 2019.

Horvath, Esther, Sebastian Grote, and Katharina Weiss-Tuider. *Into the Arctic Ice: The Largest Polar Expedition of All Time*. Munich: Prestel, 2020.

Lopez, Barry. *Arctic Dreams: Imagination and Desire in a Northern Landscape*. New York: Charles Scribner's Sons, 1986.

Mulvaney, Kieran. *At the Ends of the Earth: A History of the Polar Regions*. Washington, DC: Island Press, 2001.

Mulvaney, Kieran. *The Great White Bear: A Natural and Unnatural History of the Polar Bear*. Boston: Houghton Mifflin Harcourt, 2011.

Nuttall, Mark, ed. *Encyclopedia of the Arctic*. 3 vols. New York: Routledge, 2005.

Nuttall, Mark, Torben R. Christensen, and Martin J. Siegert, eds. *The Routledge Handbook of the Polar Regions*. New York: Routledge, 2019.

Sale, Richard. *The Arctic: The Complete Story*. London: Frances Lincoln, 2008.

Sale, Richard. *A Complete Guide to Arctic Wildlife*. Buffalo, NY: Firefly Books, 2006.

Swaney, Deanna. *The Arctic*. Hawthorn, Victoria: Lonely Planet, 1999.

Thomas, David N., ed. *Arctic Ecology*. Oxford: Wiley Blackwell, 2021.

Polar Peoples and Perspectives

Brody, Hugh. *Living Arctic: Hunters of the Canadian North*. Seattle: University of Washington Press, 1987.

Damas, David, ed. *Handbook of North American Indians*. Vol. 5, *Arctic*. Washington, DC: Smithsonian Institution Press, 1984.

Dumond, Don E. *The Eskimos and Aleuts*. London: Thames & Hudson, 1977.

Eber, Dorothy Harley. *Encounters on the Passage: Inuit Meet the Explorers*. Toronto: University of Toronto Press, 2008.

Fossett, Renée. *In Order to Live Untroubled: Inuit of the Central Arctic, 1550 to 1940*. Winnipeg: University of Manitoba Press, 2001.

Gearheard, Shari Fox, Lene Kielsen Holm, Henry Huntington, Joe Mello Leavitt, Andrew R. Mahoney, Margaret Opie, Toku Oshima, and Joelie Sanguya, eds. *The Meaning of Ice: People and Sea Ice in Three Arctic Communities.* Hanover, NH: International Polar Institute Press, 2013.

Kenney, Gerard I. *Arctic Smoke & Mirrors.* Prescott, ON: Voyageur, 1994.

Malaurie, Jean. *Ultima Thule: Explorers and Natives in the Polar North.* Translated from the French by Willard Wood and Anthony Roberts. New York: W. W. Norton, 2003.

Marcus, Alan R. *Out in the Cold: The Legacy of Canada's Inuit Relocation Experiment in the High Arctic.* Copenhagen: International Work Group for Indigenous Affairs, 1992.

McGhee, Robert. *Ancient People of the Arctic.* Vancouver: UBC Press, 1996.

McGrath, Melanie. *The Long Exile: A Tale of Inuit Betrayal and Survival in the High Arctic.* New York: Vintage Books, 2008.

Minority Rights Group. *Polar Peoples: Self-Determination and Development.* London: Minority Rights Publications, 1994.

Watt-Cloutier, Sheila. *The Right to Be Cold: One Woman's Fight to Protect the Arctic and Save the Planet from Climate Change.* Minneapolis: University of Minnesota Press, 2018.

Wright, Shelley. *Our Ice Is Vanishing / Sikuvut Nunguliqtuq: A History of Inuit, Newcomers, and Climate Change.* Montreal: McGill-Queen's University Press, 2014.

Exploration and History

Amundsen, Roald. *My Life as an Explorer.* Garden City, NY: Doubleday, Doran, 1928.

Amundsen, Roald. *The North West Passage.* New York: E. P. Dutton, 1908.

Barrow, John. *A Chronological History of Voyages into the Arctic Regions: Undertaken Chiefly for the Purpose of Discovering a North-East, North-West, or Polar Passage between the Atlantic and Pacific.* London: John Murray, 1818. Reprint, Newton Abbott, UK: David & Charles, 1971.

Berton, Pierre. *The Arctic Grail: The Quest for the Northwest Passage and the North Pole, 1818–1909.* New York: Lyons Press, 2001.

Delgado, James P. *Across the Top of the World: The Quest for the Northwest Passage.* New York: Checkmark Books, 1999.

Fleming, Fergus. *Barrow's Boys: The Original Extreme Adventurers.* New York: Atlantic Monthly Press, 1998.

Holland, Clive. *Arctic Exploration and Development, c. 500 B.C. to 1915.* New York: Garland, 1994.

Huntford, Roland. *The Last Place on Earth: Scott and Amundsen's Race to the South Pole.* New York: Atheneum, 1985.

Kish, George. *North-east Passage: Adolf Erik Nordenskiöld, His Life and Times.* Amsterdam: Nico Israel, 1973.

McGoogan, Ken. *Dead Reckoning: The Untold Story of the Northwest Passage.* New York: HarperCollins, 2017.

Mirsky, Jeannette. *To the Arctic! The Story of Northern Exploration from Earliest Times.* Chicago: University of Chicago Press, 1970.

Palin, Michael. *Erebus: One Ship, Two Epic Voyages, and the Greatest Naval Mystery of All Time.* Vancouver, BC: Greystone Books, 2018.

Potter, Russell A. *Finding Franklin: The Untold Story of a 165-Year Search.* Montreal: McGill-Queen's University Press, 2016.

Robinson, Michael F. *The Coldest Crucible: Arctic Exploration and American Culture.* Chicago: University of Chicago Press, 2006.

Savours, Ann. *The Search for the North West Passage.* New York: St. Martin's Press, 1999.

Vaughan, Richard. *The Arctic: A History.* Dover, NH: Alan Sutton, 1994.

Watson, Paul. *Ice Ghosts: The Epic Hunt for the Lost Franklin Expedition.* New York: W. W. Norton, 2017.

Williams, Glyn. *Arctic Labyrinth: The Quest for the Northwest Passage.* Berkeley: University of California Press, 2010.

Woodman, David C. *Unravelling the Franklin Mystery: Inuit Testimony.* Montreal: McGill-Queen's University Press, 2015.

Zaslow, Morris, ed. *A Century of Canada's Arctic Islands: 1880–1980.* Ottawa: Royal Society of Canada, 1981.

Geopolitics

Brady, Anne-Marie. *China as a Polar Great Power.* Cambridge, UK: Cambridge University Press, 2017.

Breum, Martin. *Cold Rush: The Astonishing True Story of the New Quest for the Polar North.* Montreal: McGill-Queen's University Press, 2018.

Brigham, Lawson W., ed. *The Soviet Maritime Arctic.* Annapolis, MD: Naval Institute Press, 1991.

Buchanan, Elizabeth. *Red Arctic: Russian Strategy under Putin.* Washington, DC: Brookings Institution Press, 2023.

Byers, Michael. *Who Owns the Arctic? Understanding Sovereignty Disputes in the North*. Vancouver, BC: Douglas & McIntyre, 2009.

Coates, Ken S., and Carin Holroyd, eds. *The Palgrave Handbook of Arctic Policy and Politics*. Cham, Switzerland: Palgrave Macmillan, 2020.

Coates, Ken S., P. Whitney Lackenbauer, William R. Morrison, and Greg Poelzer. *Arctic Front: Defending Canada in the Far North*. Toronto: Thomas Allen, 2008.

Dodds, Klaus, and Mark Nuttall. *The Scramble for the Poles: The Geopolitics of the Arctic and Antarctic*. Cambridge, UK: Polity Press, 2016.

Gjørv, Gunhild Hoogensen, Marc Lanteigne, and Horatio Sam-Aggrey, eds. *Routledge Handbook of Arctic Security*. New York: Routledge, 2022.

Grant, Shelagh D. *Polar Imperative: A History of Arctic Sovereignty in North America*. Vancouver, BC: Douglas & McIntyre, 2010.

Heininen, Lassi, and Heather Exner-Pirot, eds. *Climate Change and Arctic Security: Searching for a Paradigm Shift*. Cham, Switzerland: Palgrave Macmillan, 2020.

Honderich, John. *Arctic Imperative: Is Canada Losing the North?* Toronto: University of Toronto Press, 1987.

Hønneland, Geir. *Russia and the Arctic: Environment, Identity and Foreign Policy*. London: I.B. Tauris, 2020.

Spohr, Kristina, and Daniel S. Hamilton, eds. *The Arctic and World Order*. Washington, DC: Johns Hopkins University, 2020.

Tamnes, Rolf, and Kristine Offerdal, eds. *Geopolitics and Security in the Arctic: Regional Dynamics in a Global World*. New York: Routledge, 2016.

Sea Ice, Climate Change, and Other Environmental Issues

Anderson, Alun. *After the Ice: Life, Death, and Geopolitics in the New Arctic*. New York: Smithsonian Books, 2009.

Ehrlich, Gretel. *The Future of Ice: A Journey into Cold*. New York: Pantheon Books, 2004.

Emmerson, Charles. *The Future History of the Arctic*. New York: PublicAffairs, 2010.

Gornitz, Vivien. *Vanishing Ice: Glaciers, Ice Sheets, and Rising Seas*. New York: Columbia University Press, 2019.

Gosnell, Mariana. *Ice: The Nature, the History, and the Uses of an Astonishing Substance*. New York: Alfred A. Knopf, 2005.

Jamail, Dahr. *The End of Ice: Bearing Witness and Finding Meaning in the Path of Climate Disruption*. New York: New Press, 2019.

Johannessen, Ola M., Leonid P. Bobylev, Elena V. Shalina, and Stein Sandven, eds. *Sea Ice in the Arctic: Past, Present and Future*. Cham, Switzerland: Springer, 2020.

Serreze, Mark C. *Brave New Arctic: The Untold Story of the Melting North*. Princeton, NJ: Princeton University Press, 2018.

Struzik, Edward. *Future Arctic: Field Notes from a World on the Edge*. Washington, DC: Island Press, 2015.

Thomas, David N. *Frozen Oceans: The Floating World of Pack Ice*. Buffalo, NY: Firefly Books, 2004.

Wadhams, Peter. *Ice in the Ocean*. London: Gordon and Breach Science Publishers, 2000.

Weart, Spencer R. *The Discovery of Global Warming*. Cambridge, MA: Harvard University Press, 2003.

About the Author

Kieran Mulvaney is the author of *At the Ends of the Earth: A History of the Polar Regions, The Whaling Season: An Inside Account of the Struggle to Stop Commercial Whaling,* and *The Great White Bear: A Natural and Unnatural History of the Polar Bear.* A regular contributor to *National Geographic,* he  has also written for the *Guardian,* the *Washington Post Magazine, BBC Wildlife, New Scientist, E Magazine,* and other publications.

Born in England, he spent several years living in a cabin in Alaska and visits the Arctic and subarctic regularly. He now lives in rural Vermont.

Index

Maps and photos are in italics.